Lecture Notes in Computer Science 13650

Founding Editors

Gerhard Goos
  Karlsruhe Institute of Technology, Karlsruhe, Germany
Juris Hartmanis
  Cornell University, Ithaca, NY, USA

Editorial Board Members

Elisa Bertino
  Purdue University, West Lafayette, IN, USA
Wen Gao
  Peking University, Beijing, China
Bernhard Steffen ⓘD
  TU Dortmund University, Dortmund, Germany
Moti Yung ⓘD
  Columbia University, New York, NY, USA

More information about this series at https://link.springer.com/bookseries/558

Renata Guizzardi · Bernd Neumayr (Eds.)

Advances in Conceptual Modeling

ER 2022 Workshops, CMLS, EmpER, and JUSMOD
Hyderabad, India, October 17–20, 2022
Proceedings

Springer

Editors
Renata Guizzardi (iD)
University of Twente
Enschede, The Netherlands

Bernd Neumayr (iD)
Johannes Kepler University Linz
Linz, Austria

ISSN 0302-9743   ISSN 1611-3349 (electronic)
Lecture Notes in Computer Science
ISBN 978-3-031-22035-7   ISBN 978-3-031-22036-4 (eBook)
https://doi.org/10.1007/978-3-031-22036-4

This Springer imprint is published by the registered company Springer Nature Switzerland AG
The registered company address is: Gewerbestrasse 11, 6330 Cham, Switzerland

Preface

The International Conference on Conceptual Modeling (ER) is the leading conference on conceptual modeling, along with its foundations and applications. This year, the 41st edition of ER was held online from October 17 to 20, 2022. In addition to its main program, the conference hosted several satellite events, including four workshops.

ER workshops are forums for exchanging emergent ideas on conceptual modeling. They provide opportunities for presenting initial and ongoing work and allow researchers and practitioners to interact and collaborate. This volume contains the proceedings of three of the four workshops held in conjunction with ER 2022. The Program Committees (PCs) and chairs of the workshops were responsible for reviewing and selecting the papers. Each paper was reviewed single-blind by at least three members of the respective Program Committee. Out of 20 submissions, the PCs accepted 11 papers for publication in these proceedings.

Anna Bernasconi, Arif Canakoglu, Ana León Palacio, and José Fabián Reyes Román organized the 3rd International Workshop on Conceptual Modeling for Life Sciences (CMLS 2022). The CMLS workshop aims at bringing together researchers from three different areas, namely information systems, conceptual modeling, and data management, to share their work on problems in healthcare and life sciences, with a focus on genomic data management and precision medicine. This volume contains the five papers presented at CMLS 2022.

João Araujo, Dominik Bork, Miguel Goulão, and Sotirios Liaskos organized the 5th International Workshop on Empirical Methods in Conceptual Modeling (EmpER 2022). EmpER brings together researchers and practitioners interested in the empirical investigation of conceptual modeling systems and practices. This volume contains the one paper presented at EmpER 2022. This year, CMLS and EmpER merged their program into two joint sessions featuring a keynote speech by Pietro Pinoli from the Politecnico di Milano.

Silvana Castano, Mattia Falduti, Cristine Griffo, Stefano Montanelli, and Tiago Prince Sales organized the 1st International Workshop on Digital Justice, Digital Law, and Conceptual Modeling (JUSMOD 2022). JUSMOD invites researchers to share ideas on modeling, analyzing, formalizing, and interpreting legal data and related processes. This volume contains the five papers presented at JUSMOD 2022. The workshop also featured a keynote speech by Enrico Francesconi, entitled "Legal Knowledge Representation and Reasoning in the Semantic Web".

Tong Li, Alejandro Mate, and Enyo Gonçalves organized the 15th International i* workshop (iStar 2022). The i* workshop brings together researchers in the area of goal modeling to exchange ideas, compare notes, promote interactions, and forge new collaborations. The eight papers presented at iStar 2022 are published in a volume of CEUR Workshop Proceedings.

We are deeply grateful to the ER 2022 general chairs, Kamalakar Karlapalem and Manfred Jeusfeld, and PC chairs, Upendranath Chakravarthy, Mukesh Mohania, and Jolita Ralyté, for their trust and support. We thank the workshop organizers for organizing the workshops and putting together exciting programs composed of keynote speeches, presentations of technical papers, and discussions. We are grateful to the workshop authors and reviewers for ensuring the high quality of the workshop programs. Finally, we thank the team at Springer for their guidance and help in preparing these proceedings.

October 2022   Renata Guizzardi
                                                        Bernd Neumayr

ER 2022 Conference Organization

General Chairs

Kamalakar Karlapalem   IIIT Hyderabad, India
Manfred Jeusfeld   University of Skövde, Sweden

Program Committee Chairs

Upendranath Chakravarthy   University of Texas at Arlington, USA
Mukesh Mohania   IIIT Delhi, India
Jolita Ralyté   University of Geneva, Switzerland

Workshop Chairs

Renata Guizzardi   University of Twente, The Netherlands
Bernd Neumayr   Johannes Kepler University Linz, Austria

Tutorial Chairs

Hans-Georg Fill   University of Fribourg, Switzerland
Vinay Kulkarni   Tata Consultancy Services, India

Panel Chairs

Vikram Goyal   IIIT Delhi, India
Carson Woo   University of British Columbia, Canada

Forum, Demo, and Posters Chairs

Sebastian Link   University of Auckland, New Zealand
Iris Reinhartz-Berger   University of Haifa, Israel
Jelena Zdravkovic   Stockholm University, Sweden

Sponsoring and Industry Chairs

Aditya Ghose   University of Wollongong, Australia
Henderik Proper   Luxembourg Institute of Science and Technology, Luxembourg

PhD Symposium Chairs

Srinath Srinivasa	IIIT Bangalore, India
Dominik Bork	TU Wien, Austria

Publicity and Social Media Chairs

Judith Michael	RWTH Aachen University, Germany
Marcela Ruiz	Zürich University of Applied Sciences, Switzerland

Local Organizing Chair

P. Radha Krishna	National Institute of Technology Warangal, India

Web Chairs

Sangharatna Godboley	National Institute of Technology Warangal, India
Syed Juned Ali	TU Wien, Austria

ER 2022 Workshops Organization

Conceptual Modeling for Life Sciences (CMLS)

Organizing Co-chairs

Anna Bernasconi	Politecnico di Milano, Italy
Arif Canakoglu	Fondazione IRCCS Ca' Granda Ospedale Maggiore Policlinico, Italy
Ana León Palacio	Universitat Politècnica de València, Spain
José Fabián Reyes Roman	Universitat Politècnica de València, Spain

Program Committee

Giuseppe Agapito	Magna Graecia University, Italy
Samuele Bovo	University of Bologna, Italy
Bernardo Breve	University of Salerno, Italy
Mario Cannataro	Magna Graecia University, Italy
Stefano Cirillo	University of Salerno, Italy
Johann Eder	University of Klagenfurt, Germany
Jose Luis Garrido	University of Granada, Spain
Giancarlo Guizzardi	University of Twente, The Netherlands
Khanh N. Q. Le	Taipei Medical University, Taiwan
Sergio Lifschitz	Pontifical Catholic University of Rio de Janeiro, Brazil
Paolo Missier	Newcastle University, UK
José Palazzo	Federal University of Rio Grande do Sul, Brazil
Ignacio Panach	University of Valencia, Spain
Barbara Pernici	Polytechnic University of Milan, Italy
Rosario Michael Piro	Polytechnic University of Milan, Italy
Monjoy Saha	National Cancer Institute, USA
Domenico Vito	University of Pavia, Italy
Emanuel Weitschek	UniNettuno University, Italy

Empirical Methods in Conceptual Modeling (EmpER)

Organizing Co-chairs

João Araujo	Universidade Nova de Lisboa, Portugal
Dominik Bork	TU Wien, Austria
Miguel Goulão	Universidade Nova de Lisboa, Portugal
Sotirios Liaskos	York University, Canada

Program Committee

Robert Buchmann	Babeș-Bolyai University, Romania
Jordi Cabot	Open University of Catalonia, Spain
Michel Chaudron	Eindhoven University of Technology, The Netherlands
Neil Ernst	University of Victoria, Canada
Marcela Fabiana Genero Bocco	University of Castilla-La Mancha, Spain
Antonio Garmendia	Johannes Kepler University Linz, Austria
Sepideh Ghanavati	University of Maine, USA
Andrea Herrmann	Heidelberg University, Germany
Katsiaryna Labunets	Utrecht University, The Netherlands
Lin Liu	Tsinghua University, China
Jan Mendling	Humboldt-Universität zu Berlin, Germany
Raimundas Matulevičius	University of Tartu, Estonia
Ana Moreira	Universidade Nova de Lisboa, Portugal,
Xavier Le Pallec	University of Lille 1, France
Jeffrey Parsons	Memorial University, Canada
Oscar Pastor	Universidad Politécnica de Valencia, Spain
Geert Poels	Ghent University, Belgium
Jan Recker	University of Hamburg, Germany
Ben Roelens	Open Universiteit, Belgium
Carla Silva	Universidade Federal de Pernambuco, Brazil
Manuel Wimmer	Johannes Kepler University Linz, Austria

Digital Justice, Digital Law and Conceptual Modeling (JUSMOD)

Organizing Co-chairs

Silvana Castano	Università degli Studi di Milano, Italy
Mattia Falduti	Free University of Bozen-Bolzano, Italy
Cristine Griffo	Free University of Bozen-Bolzano, Italy
Stefano Montanelli	Università degli Studi di Milano, Italy
Tiago Prince Sales	University of Enschede, The Netherlands

Program Committee

Tommaso Agnoloni	IGSG-CNR, Italy
João Paulo A. Andrade	Federal University of Espírito Santo, Brazil
Lauro Araujo	Prodasen - Center for Data Processing of the Federal Senate, Brazil
Jean-Rémi Bourguet	University of Vila Velha, Brazil
Samuel M. Brasil Jr.	National School of the Judiciary (ENFAM), Brazil
Daniel Braun	University of Twente, The Netherlands
Mirna El Ghosh	INSA Rouen Normandie, France
Matthias Grabmair	Technische Universität München, Germany
Giancarlo Guizzardi	University of Twente, The Netherlands
Juliana Justo Castello	Vitória Law School, Brazil
Juliano Maranhão	São Paulo University, Brazil
Samuela Marchiori	University of Twente, The Netherlands
Robert Muthuri	Strathmore University, Kenya
Elias Oliveira	Federal University of Espírito Santo, Brazil
Matteo Palmonari	Università di Milano-Bicocca, Italy
Anca Radu	European University Institute (EUI), Italy
Aires Rover	Federal University of Santa Catarina, Brazil
Jaromir Savelka	Carnegie Mellon University, USA
Giovanni Sileno	University of Amsterdam, The Netherlands
Paolo Sommaggio	University of Trento, Italy
Andrea Tagarelli	Università della Calabria, Italy
Maria das Graças Teixeira	Federal University of Espírito Santo, Brazil
Sergio Tessaris	Free University of Bozen-Bolzano, Italy
Ester Zumpano	Università della Calabria, Italy

Contents

CMLS and EmpER

Towards a Model-Driven Approach for Big Data Analytics in the Genomics
Field 5
    *Ana Xavier Fernandes, Filipa Ferreira, Ana León,
    and Maribel Yasmina Santos*

Conceptual Modeling-Based Cardiopathies Data Management 15
    Mireia Costa, Alberto García S., and Oscar Pastor

A Conceptual Model of Health Monitoring Systems Centered on ADLs
Performance in Older Adults 25
    *Francisco M. Garcia-Moreno, Mãria Bermudez-Edo,
    José Manuel Pérez Mármol, José Luis Garrido,
    and María José Rodríguez-Fórtiz*

A Comparative Analysis of the Completeness and Concordance of Data
Sources with Cancer-Associated Information 35
    Mireia Costa, Alberto García S., and Oscar Pastor

A Flexible Automated Pipeline Engine for Transcript-Level Quantification
from RNA-seq 45
    Pietro Cinaglia and Mario Cannataro

An Initial Empirical Assessment of an Ontological Model of the Human
Genome 55
    *Alberto García S., Anna Bernasconi, Giancarlo Guizzardi,
    Oscar Pastor, Veda C. Storey, and Mireia Costa*

JUSMOD

Unsupervised Factor Extraction from Pretrial Detention Decisions
by Italian and Brazilian Supreme Courts 69
    *Isabela Cristina Sabo, Marco Billi, Francesca Lagioia,
    Giovanni Sartor, and Aires José Rover*

Context-Aware Knowledge Extraction from Legal Documents Through
Zero-Shot Classification 81
    Alfio Ferrara, Sergio Picascia, and Davide Riva

On the Lack of Legal Regulation in Conceptual Modeling 91
    Kai von Lewinski and Stefanie Scherzinger

Automated Extraction and Representation of Citation Network: A CJEU
Case-Study 102
    *Galileo Sartor, Piera Santin, Davide Audrito, Emilio Sulis,
    and Luigi Di Caro*

A Rule 74 for Italian Judges and Lawyers 112
    Giulia Pinotti, Amedeo Santosuosso, and Federica Fazio

Author Index 123

CMLS and EmpER

3rd International Workshop on Conceptual Modeling for Life Sciences (CMLS 2022) and 5th International Workshop on Empirical Methods in Conceptual Modeling (EmpER 2022)

Anna Bernasconi[1], Arif Canakoglu[2(✉)], Ana León Palacio[3],
José Fabián Reyes Román[3], João Araujo[4], Dominik Bork[5],
Miguel Goulão[4], and Sotirios Liaskos[6]

[1] Dipartimento di Elettronica, Informazione e Bioingegneria, Politecnico di
Milano, Milan, Italy
anna.bernasconi@polimi.it

[2] Dipartimento di Anestesia, Rianimazione ed Emergenza-Urgenza, Fondazione
IRCCS Cà Granda Ospedale Maggiore Policlinico, Milan, Italy
arif.canakoglu@policlinico.mi.it

[3] Valencian Research Institute for Artificial Intelligence (VRAIN), Universitat
Politècnica de València, Valencia, Spain
{aleon,jreyes}@vrain.upv.es

[4] Department of Informatics, Universidade Nova de Lisboa, Lisbon, Portugal
{pl91,mgoul}@fct.unl.pt

[5] Business Informatics Group, TU Wien, Vienna, Austria
dominik.bork@tuwien.ac.at

[6] School of Information Technology, York University, Toronto, Canada
liaskos@yorku.ca

Exceptionally for ER 2022, the 3rd International Workshop on Conceptual Modeling for Life Sciences (CMLS 2022) is joined by the 5th International Workshop on Empirical Methods in Conceptual Modeling (EmpER 2022).

The recent advances in unraveling the secrets of human conditions and diseases have encouraged new paradigms for their prevention, diagnosis, and treatment. As information is increasing at an unprecedented rate, it directly impacts the design and future development of information and data management pipelines; thus, new ways of processing data, information, and knowledge in health care environments are strongly needed. The objective of CMLS 2022 is to continue being a meeting point for Information Systems (IS), Conceptual Modeling (CM), and Data Management (DM) researchers working on health care and life science problems. It is also an opportunity to share, discuss and find new approaches to improve promising fields, with a special focus on *Genomic Data Management* – how to use the information from the genome to better understand biological and clinical features – and *Precision Medicine* – giving to each patient an individualized treatment by understanding the peculiar aspects of the disease. The joined research communities of IS, CM, and DM have an important role to play; they must help in providing feasible solutions for high-quality and efficient health care.

For the 2022 edition, CMLS is joined by EmpER 2022, which aimed at bringing together researchers with an interest in the empirical investigation of conceptual modeling. Like its predecessors, this edition of the workshop invited three kinds of papers: *full study papers* describing a completed empirical study, *work-in-progress papers* describing a planned study or study in progress, as well as *position, vision and lessons learned papers* about the use of empirical methods for conceptual modeling. The workshop is friendly to negative results as well as proposed empirical studies that are in their design stage so that authors can benefit from early feedback. A total of twenty reviewers were invited to serve the program committee of the workshop based on their past contributions in the area of empirical conceptual modeling.

The CMLS 2022 and EmpER 2022 workshops handled the review process separately. The third edition of CMLS has attracted high-quality submissions about models of the life sciences domain. Five papers have been selected after a blind review process that involved three experts from the field for each submission. The topics addressed by the accepted papers concern big data analytics for genomics, conceptual modeling for data management of cardiopathies and cancer-related information, RNA-quantification, and health monitoring software. All of them confirm an interesting technical program to stimulate live discussion. We expect a growing interest in this area in the coming years; this was one of the motivations for continuing our commitment to this workshop in conjunction with the ER 2022 conference. The five papers from CMLS were joined by one EmpER 2022 paper describing a completed empirical study on the effectiveness of an ontology-oriented conceptual modeling language (OntoUML) for modeling aspects of the human genome.

The two workshops will be run in a common session and interdisciplinary discussion will be promoted, inviting exchange between the communities of the two workshops.

Acknowledgements. CMLS 2022 was organized within the framework of the projects Advanced Grant 693174 (data-driven Genomic Computing) – European Research Council, PDC2021-121243-I00 (DELFOS) – Spanish State Research Agency, INNEST/2021/57 (OGMIOS) – Agència Valenciana de la Innovacièn, Generalitat Valenciana. We would like to express our gratitude to Stefano Ceri and Oscar Pastor who demonstrated continuous support to the organization of CMLS 2022. We thank the Program Committee members for their hard work in reviewing papers, the authors for submitting their works, and the ER 2022 organizing committee for supporting our workshop. We also thank ER 2022 workshop chairs Renata S. S. Guizzardi and Bernd Neumayr for their direction and guidance.

Towards a Model-Driven Approach for Big Data Analytics in the Genomics Field

Ana Xavier Fernandes[1]([✉]) [iD], Filipa Ferreira[1] [iD], Ana León[1,2] [iD],
and Maribel Yasmina Santos[1] [iD]

[1] ALGORITMI Research Centre, University of Minho, Guimarães, Portugal
{a85638,a78447}@alunos.uminho.pt, maribel@dsi.uminho.pt
[2] Valencian Research Institute for Artificial Intelligence (VRAIN), Universitat
Politècnica de València, Camí de Vera S/N, Valencia, Spain
aleon@vrain.upv.es

Abstract. The use of techniques such as Next Generation Sequencing
has allowed a fast increase in data generation due to the reduction of processing costs. What at the beginning seemed to be an important step forward for the development of new approaches such as Precision Medicine,
turned into an exponential growth of data that currently challenges
healthcare professionals and researchers. Since the problems derived from
the storage and management of vast amounts of heterogeneous data are
well-known for the Big Data and Information Systems communities, the
application of this knowledge to the genomic data domain can help to
improve the management of the data, reduce the bottlenecks, and reveal
new insights on the causes of human disease. In this way, this work is
focused on the problem of data storage by proposing a Big Data architecture supported by a model-driven approach to ensure an efficient and
dynamic storage of genomic data. The proposed architecture has been
designed considering the main requirements for an efficient data integration and for supporting data analysis tasks.

Keywords: ETL · Data integration · Data storage · Data analysis

1 Introduction

The exponential growth of genomic data derived from the application of Next
Generation Sequencing (NGS) technologies has made their management difficult,
posing new challenges for researchers and clinicians when establishing relevant
and reliable relationships between DNA variants and genetic diseases. Currently,
over 1,000 public repositories are available and often used by the scientific community. Nevertheless, these repositories are characterized by their heterogeneity
and diverse level of quality [1], constituting the first challenge to face. The second challenge is the dispersion, because there are hundreds of different relevant
genomic databases, many of them created and removed every year. Finally, the

R. Guizzardi and B. Neumayr (Eds.): ER 2022 Workshops, LNCS 13650, pp. 5–14, 2022.
https://doi.org/10.1007/978-3-031-22036-4_1

third challenge is the lack of interoperability since all these data are difficult to integrate and interconnect due to the two previously mentioned problems.

To face these challenges, appropriate approaches and technologies are needed to support data storage, integration and analysis, reducing the cost of such tasks that currently remain highly manual [2]. Furthermore, the integration of new data sources to the system must also be considered, specially in highly dynamic domains like Genomics. Since all these challenges are well-known for the Big Data and Information Systems communities, the application of this knowledge to the genomic domain can help to improve the management of the data, reducing the bottlenecks, and helping to reveal new insights on the causes of human diseases.

A first step on this direction was the SILE method [3], which establishes bases for selecting the most appropriate genomic repositories (S), identifying the relevant information according to the analytical requirements (I), ensuring data persistence (L), and developing tools to exploit the data (E). The SILE method is supported by a technological solution (the Delfos platform) made up of four modules (Hermes, Ulises, Delfos and Sibila) that provides automation to each of the main steps of the method.

This work is focused on proposing the Big Data storage architecture that supports the Hermes and Delfos modules, as basis to build an efficient data management and analysis system. The proposed architecture is supported by a model-driven approach to ensure an efficient and dynamic data storage. This paper is focused on the Hermes and Delfos modules. They support the extraction, cleaning, transformation, and integration of data obtained from the genomic repositories (Hermes), and the storage of the data for further analysis (Delfos).

To such aim, this paper is structured as follows. Section 2 summarizes the work related to ETL (Extraction-Transformation-Loading) processes, the integration of genomic data, as well as their storage and analysis. Section 3 proposes the Big Data Architecture that supports the Delfos platform, detailing two of its main modules: the Hermes module and the Delfos module. Section 4 discusses the obtained results and concludes with proposals for future work.

2   Related Work

Advances in sequencing technologies and data processing pipelines are rapidly providing sequencing data that require Big Data computing strategies with levels of abstraction that exceed the capabilities of available tools [9]. These strategies must be focused mainly on ETL processess, data storage and data analysis, as well as on efficient data pre-pocessing and querying. Current approaches for the implementation of ETL processess has been discussed from many points of view. One attempt was centered on optimizing the physical design of ETL through a set of algorithms [4]. A formalization of the logical data storage structure and the hardware and software configurations of the ETL using UML (Unified Modeling Language) has been introduced by [5]. Although both works deal with interesting topics in the area, they do not produce code for the execution of ETL processes. An approach to ETL programming using the Python language has been devised by [6]. However, this approach omits the provision of a vendor-independent design, which decreases the re-usability and ease of use of the provided framework. In the field of Big Data Analytics, an intelligent system is proposed in the work of [7] that aims to provide effective support to companies based on a process of research and selection of artefacts of interest from the Web, in order to verify how they coincide with the experience and expertise of the company, stored in its internal data and knowledge sources. However, this study only covers the domain of E-Procurement. Regarding data storage and analysis, in [8] the authors presents a Big Data approach with the intention of improving genetic diagnosis in the context of Precision Medicine. The study has proven to be beneficial for a better evaluation of genetic diagnosis of complex diseases such as epilepsy. Regarding query languages, in [9] the authors propose the GenoMetric Query Language (GMQL) for raw data pre-processing and querying over diverse datasets. This language is supported by the Hadoop platform and Apache Pig, which ensures scalability, expressiveness, flexibility and simplicity of use. In the mentioned related works, the proposed storage and analysis systems use significantly smaller genomic data samples. Additionally, these systems use various technologies and tools that require experts because the code must be configured to make possible changes to the system. The work presented in this paper overcomes these limitations by following a Model-Driven approach that allows abstracting all the complexity of the data pipelines needed for the ETL processes, as well as the data analysis tasks.

3   Delfos Architecture and Implementation

The architecture of the Delfos platform (Fig. 1) includes four modules (Hermes, Delfos, Ulises and Sibila), and three additional components (the Genomic Raw Data component, the Conceptual Schema, and the Domain Expert component).

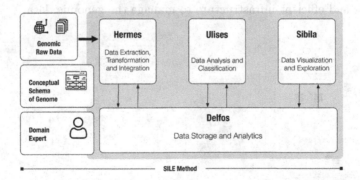

Fig. 1. Logical architecture of the Delfos platform

The Hermes module is responsible for the extraction, transformation and integration of data. The Delfos module is transversal to the whole platform, supporting the storage and analysis of the data. The Ulises module aims to apply machine learning algorithms to the integrated and treated data, while Sibila is responsible not only for exploring, but also for visualizing the stored data. The tasks automated by these modules are derived from the SILE method.

The Genomic Raw Data component comprises the sources that provide the genomic data to the system. These sources can be local or external data sources. The platform is supported by the Conceptual Schema of the Genome [10] [11], that provides an accurate representation of the relevant domain concepts. The Domain Expert has the specific knowledge about the source data, how it should be extracted from the data sources, and which transformations should be applied.

In terms of the data flow that takes place between the components, the Domain Expert establishes the communication between the genomic data sources, in XML or CSV format, and the Hermes module, which loads the raw data into the Delfos module to perform the ETL tasks. During this process, Hermes communicates with the Delfos module to store the temporary data to be used during the processing. After processing and storing the data in JSON and CSV formats, the prepared and treated data can be used by the other modules (Ulises and Sibila). The Ulises module can apply machine learning algorithms and further store them in the Delfos module. The Sibila module interacts with the Delfos module through a unidirectional arrow as it uses the prepared data provided by the Delfos module to create dashboards and other visualizations. This work is focused on describing in more detail the Hermes and Delfos modules, making the platform flexible and efficient.

3.1   The Hermes Module

The Hermes module integrates a logical architecture and a technological architecture (Fig. 2). The main goal of the logical architecture is to automate the tasks as much as possible in order to deal with the complexity that the genomic domain presents. The main goal of the technological architecture is to guarantee an scalable and efficient infrastructure to manage the genomic data.

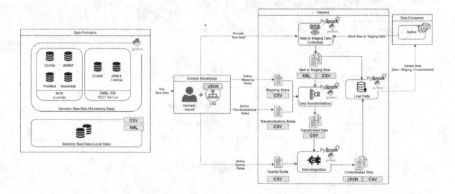

Fig. 2. Logical and technological architectures of the Hermes module

To this aim, four main components need to interact: Data Providers, Domain Knowledge, Hermes and Data Consumer. The Data Providers component represents the data sources that provide the genomic data to the system. This corresponds to the Genomic Raw Data component described in Fig. 1. Data can be extracted from local or from external data sources (such as ClinVar, Ensembl, or PubMed). Current accepted data formats are XML and CSV.

The Domain Knowledge component includes the Domain Expert and the Conceptual Schema of the Genome (CSG), that represents the main domain concepts that will be processed and used by the Delfos platform. The CSG is represented by a unified data schema, in this case a JSON schema[1]

Taking the CSG into consideration, the expert is responsible for defining a set of rules that will allow i) the extraction of the data from data sources (mapping rules), ii) the transformations that should be applied to the data (transformation rules), and iii) the data quality requirements (quality rules). The Hermes component processes the set of rules defined by the domain expert. These rules are specified in comma delimited format (CSV), and they are applied to the raw data in order to obtain a set of consolidated/processed/treated data according to the structure defined by the JSON schema.

The Hermes component integrates four other components: Raw or Staging Data Collection, Data Transformation, Data Integration, and Load Data. The first component refers to the raw data and the data that are not fully processed to be used later. These data are used by the Data Consumer and is stored in the Delfos module. The Data Transformation module receives these data, along with the mapping and the transformation rules. This module is in charge of applying the rules to extract the needed data and to make the necessary transformations to conform the specifications of the JSON schema. The transformed data (in CSV format) are all integrated into the Data Integration component, resulting in a consolidated file that can be represented in CSV or JSON formats. The quality of the consolidated data is measured by applying the quality rules defined by the Domain Expert. The Load Data component is responsible for loading all the types of data used (raw, staging and consolidated) in the Delfos module. In terms of technologies, PySpark is used as well as other Python libraries[2] such as pandas, xml.etree.ElementTree, and HdfsCLI.

As mentioned before, the Domain Expert needs to define a set of rules to extract, transform and verify the quality of the data. A set of templates were defined for each case: two templates to define the mapping rules (one for the CSV format and another one for the XML format), a template to define the transformation rules, and a template to define the quality rules. The use of templates (defined in CSV format) allows the independence of the code allowing a higher adaptability of the system to changes. Figure 3 shows an example on how the mapping, transformation and quality rules are defined.

Figure 4 shows the interface where the user can upload a local source data file, stored in the Hadoop Distributed File System (HDFS), a CSV file with

---

[1] JSON schema specification: https://json-schema.org/.

[2] Python Package Index: https://pypi.org/,.

elementName	attrName	rootName	rootAttr	XPath		ConditionValue	FixedValue	Notes		attribute	name of column
variation		ClinVarResult-Set		//VariationArchive				All the variants as children of the root node		chromosome	chr_name
variation	chromosome	VariationArchive	Chr	/InterpretedRecord/SimpleAllele/Location/SequenceLocation[1]							

Mapping rules for the XML format of local data sources / Mapping rules for the CSV format of local data

attribute	function	regular expression	replacement		attribute	type	not	condition	value	message
chromosome	extract	([0-9]+)+[a-zA-Z]			chromosome	integer	no	range	(1, 23)	The chromosome is correct

Transformation rules / Quality rules

Fig. 3. Examples of all rules applied to the chromosome attribute

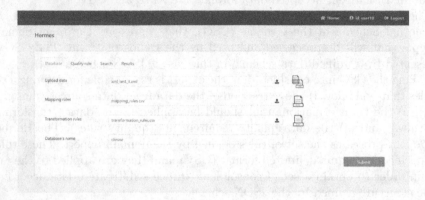

Fig. 4. Demonstration case: Submission of source data and its mapping and transformation rules

the mapping rules, and a CSV file with the transformation rules. The user can also specify a name to identify the uploaded group of files that are going to be available for querying (in this case "clinvar"). Each set of files that can be further queried is known as a job.

Figure 5 shows the interface where the user can filter the data he has submitted and use the result for further analysis. In this case, the user selects as input parameter "chromosome - variation" with the value equal to "1" in order to filter the data coming from two different sources, "clinvar" and "ensembl". It is also possible to selected the file with the quality rules to check the quality of the data after the processing. The search results can be downloaded.

For external data, the user maintains in overall the same functionalities as for the local data. In this prototype of the platform, the functionality implemented for external sources allows the search of variants that are associated to a specific phenotype. As can be seen in Fig. 6, the user selects the external database to be accessed (ClinVar) and provides the name of the phenotype (migraine with aura). The results can be filtered by any of the available parameter names (e.g., the FLNA gene).

3.2   The Delfos Module

The architecture of the Delfos module aims to optimize the storage system to become as scalable as possible, in order to meet the needs of genomics data storage and also allow an efficient data analysis. Figure 7 shows the logical and technological architecture of this module.

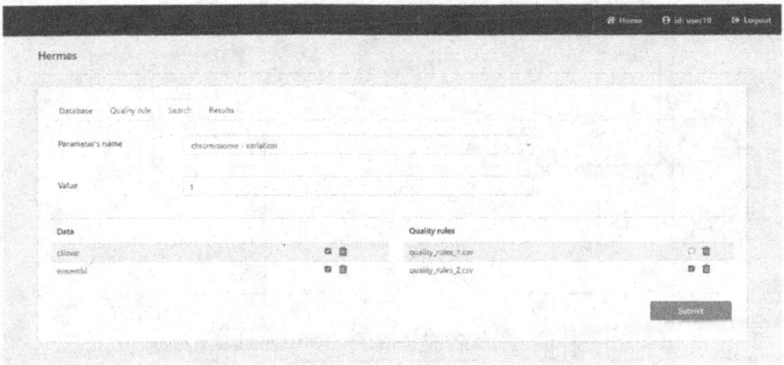

Fig. 5. Demonstration case: Performing a search

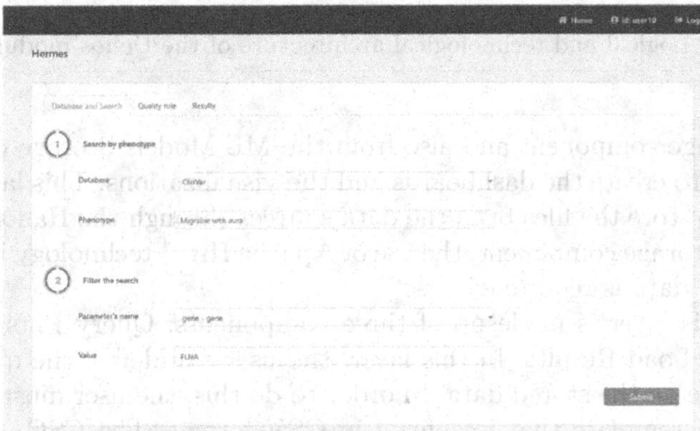

Fig. 6. Demonstration case: Database and search section

The architecture of the Delfos module is divided into two layers: the data storage layer and the data analysis layer. The storage layer has four different types of storage: Raw Data Storage, Staging Storage, Data Storage, and ML Models Storage. Raw Data Storage corresponds to the raw data that were extracted from the data sources. The Staging Storage is an area for processing, integrating, and cleaning data from temporary files. The Processed Data Storage is intended to store the data that have already been treated, processed, and integrated. Finally, the ML Models Storage aims to store the results of the Machine Learning (ML) models used in the Ulises module.

The data flow between the different storage types begins with the Hermes module that stores the raw data extracted from the data sources in the Raw Data Storage, and the temporary files in the Staging Storage component. Once the data is prepared, which means that the ETL process is completed, the data is stored in the Data Storage component. Now, these data is available for Ulises to run the ML models. The results of the models are stored in the ML Models Storage component for further use. Finally, the Sibila module reads the data from

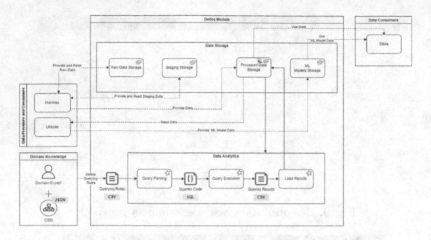

Fig. 7. Logical and technological architecture of the Delfos module

the Data Storage component and also from the ML Models Storage component and uses them to create the dashboards and the visualizations. This layer mainly uses HDFS[3] to store the files from the data sources through the Hadoop cluster. For the Data Storage component, the use of Apache Hive[4] technology is expected in case tabular data is required.

The analysis layer is made up of three components: Query Parsing, Query Execution and Load Results. In this layer, the user can define the queries that want to execute on the stored data. In order to do this, the user must define the queries using a template that is defined in CSV format. The CSG, along with the Domain Expert, provide the knowledge needed to define the query rules. The Query Parsing component is responsible for parsing the file and translating the queries to SQL language. The Query Execution component is in charge of carrying out the sequence of steps required to access the data and execute the queries. The technology used to implement these components is the Apache Spark[5] tool. Finally, the Load Results component is in charge of storing the results in the Data Storage component, that uses HDFS.

Since the architecture of the Delfos module aims to define a distributed storage layer, it is necessary to define how the file storage system will be organized. This folder hierarchy optimizes the storage of the genomic data (see Fig. 8).

The upper level of the hierarchy is the user. Going down a level in the hierarchy, it is divided into two main folders: raw_data and jobs. The raw_data folder is meant for storing data that has not been processed yet. Inside this folder, the files with the source data, the quality rules, the mapping rules, and the transformation rules are stored. The jobs folder stores all the data that have already been processed by the various modules that are part of the Delfos platform, the

---

[3] https://hadoop.apache.org/.

[4] https://hive.apache.org/.

[5] https://spark.apache.org/.

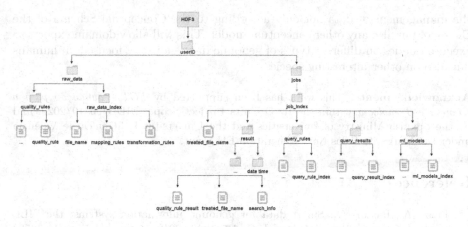

Fig. 8. Folders hierarchy

searches that have been performed, the query rules, the query results, and also the ML models.

Grouping	Operator	Parameter	Parameter type	Condition	Value
no		variant_id	string	=	VCV000021481
yes	and	hgvs	string	like	LRG%

Fig. 9. Example of query rules

Though a web interface, the user can choose the data to be queried, and upload the query rules file by selecting the option "new query". Figure 9 shows an example of how the queries are defined. After the query rules are applied, the results are stored in the HDFS and displayed to the user.

4   Discussion and Conclusions

This paper describes the architecture of the Delfos platform as a technological support for the management of vast amounts of genomic data, reducing the time required by experts to analyze such data and extract meaningful conclusions. The work is focused on the modules in charge of providing support to the extraction, transformation, integration, storage, and analysis of genomic data (Hermes and Delfos). The platform architecture has been designed following a Model-Driven approach and considering scalability and generalization.

As future work, the platform is going to be extended to easy the addition of new external sources, as well as the management of other local genomic data formats such as JSON or TXT. Additionally, in order to make the platform as much generic as possible, the system is going to be implemented to support the management of data not only according to the Conceptual Schema of the Genome, but also any other conceptual model. This will allow domain experts to execute queries on different types of genomic data, not only focused on humans but also on other interesting species.

Acknowledgements. This work has been supported by *FCT - Fundação para a Ciência e Tecnologia* within the R&D Units Project Scope: UIDB/00319/2020, and by the Spanish Ministry of Universities and the Universitat Politècnica de València under the Margarita Salas Next Generation EU grant.

References

1. León, A., Pastor, Ó.: Smart data for genomic information systems: the SILE method. Complex Syst. Inf. Model. Q. **17**, 1–23 (2018). https://doi.org/10.7250/csimq.2018-17.01
2. Galvão, J., Leon, A., Costa, C., Santos, M.Y., López, Ó.P.: Automating data integration in adaptive and data-intensive information systems. In: Themistocleous, M., Papadaki, M., Kamal, M.M. (eds.) EMCIS 2020. LNBIP, vol. 402, pp. 20–34. Springer, Cham (2020). https://doi.org/10.1007/978-3-030-63396-7_2
3. León, A.: SILE: a method for the efficient management of smart genomic information (2019). https://riunet.upv.es/handle/10251/131698
4. Tziovara, V., Vassiliadis, P., Simitsis, A.: Deciding the physical implementation of ETL workflows. In: Proceedings of the ACM 10th International Workshop on Data Warehousing and OLAP, pp. 49–56 (2007). https://doi.org/10.1145/1317331.1317341
5. Luján-Mora, S., Trujillo, J.: Physical modeling of data warehouses using UML. In: Proceedings of the 7th ACM International Workshop on Data Warehousing and OLAP, pp. 48–57 (2004). https://doi.org/10.1145/1031763.1031772
6. Thomsen, C., Bach Pedersen, T.: pygrametl: a powerful programming framework for extract-transform-load programmers. In: Proceedings of the ACM 12th International Workshop on Data Warehousing and OLAP, pp. 49–56 (2009). https://doi.org/10.1145/1651291.1651301
7. Bifulco, I., Cirillo, S., Esposito, C., Guadagni, R., Polese, G.: An intelligent system for focused crawling from big data sources. Expert Syst. Appl. **184**, 115560 (2021). https://doi.org/10.1016/j.eswa.2021.115560
8. León, A., Pastor, O.: Enhancing precision medicine: a big data-driven approach for the management of genomic data. Big Data Res. **26**, 100253 (2021). https://doi.org/10.1016/j.bdr.2021.100253
9. Masseroli, M., et al.: GenoMetric query language: a novel approach to large-scale genomic data management. Bioinformatics **31**(12), 1881–1888 (2015). https://doi.org/10.1093/bioinformatics/btv048
10. García, A., et al.: Towards the understanding of the human genome: a holistic conceptual modeling approach. IEEE Access **8**, 197111–197123 (2020). https://doi.org/10.1109/ACCESS.2020.3034793
11. García, A., et al.: A conceptual model-based approach to improve the representation and management of omics data in precision medicine. IEEE Access **9**, 154071–154085 (2021). https://doi.org/10.1109/ACCESS.2021.3128757

Conceptual Modeling-Based Cardiopathies Data Management

Mireia Costa(✉) , Alberto García S. , and Oscar Pastor

PROS Group, Valencian Research Institute (VRAIN), Universitat Politècnica de València, Camí de Vera, s/n, 46022 València, Spain
micossan@vrain.upv.es, {algarsi3,opastor}@pros.upv.es

Abstract. Familiar cardiopathies are a group of heterogeneous disorders that affect the heart. Although their origin has a strong genetic component, most patients' variations have not been reported and are not classified. Thus, cardiologists struggle to assess whether the origin of the disease is genetic and are unable to provide an appropriate early diagnosis. Providing an enriched context for understanding the role of such variations in these disorders is crucial. This work reports the identification of the conceptual dimensions that are required to provide such context. Then, we instantiate the identified conceptual dimensions in a use case. Finally, the instantiated dimensions are integrated to provide semantic interoperability.

Keywords: Conceptual modeling · Cardiomyopathies · Knowledge representation · Data management

1 Introduction

Familiar cardiopathies, also known as *inherited cardiovascular diseases*, are a set of disorders that affect the heart. The origin of these disorders tends to be genetic, with a heterogeneous clinical evolution and strong familial component. Depending on how the heart is affected, the familiar cardiopathies are classified as *cardiomyopathies*, *channelopathies*, or *genetic aortic diseases*. In this group, cardiomyopathies, which affect the heart's muscle structure and contraction processes [10], are the most prevalent, with 1 in 500 people suffering from them [1,2]. Thus, we have focused on cardiomyopathies.

A relevant problem found by cardiologists when treating patients suffering from cardiomyopathies is that most of the patients' variations have not been previously reported and are not classified (i.e., the variations have not been determined as pathogenic or benign). As a result, it is frequent that cardiologists are unable to assess whether the origin of the disease is genetic and struggle to provide an early diagnosis to patients and their relatives.

Supported by ACIF/2021/117, INBIO2021/AP2021-05, MICIN/AEI/10.13039/501100011033, and INNEST/2021/57 grants.

Identifying relevant variations and delimiting their clinical significance becomes a problem where several dimensions play a relevant role. The associated data to these dimensions have a high degree of heterogeneity and dispersion, and integrating them is a complex task that requires a conceptual model-based approach to achieve semantic interoperability between the different data types. We base our work on the Conceptual Schema of the Human Genome (CSHG) [4], which is our ontological framework of reference. The CSHG covers several dimensions of genomics with a very detailed perspective, which makes necessary the use of a method (the ISGE method) that only selects the relevant pieces of information [3].

Here, we report our preliminary work to provide clinicians with a richer context of significant variations. This work is a collaboration with several cardiologists from two Spanish hospitals that try to improve familiar cardiomyopathies' understanding[1]. Its main contributions are: **i)** the identification of the most relevant conceptual dimensions associated with genetic cardiomyopathies, **ii)** its application to a practical case through the corresponding dimensions' instantiation, and **iii)** the integration of these dimensions.

In the paper, Sect. 2 identifies the most relevant dimensions associated with genetic cardiomyopathies using the ISGE method, generating a view of the CSHG that contains a subset of relevant concepts tailored to our use case. Section 3 instantiates the identified dimensions in order to increase problem understanding. Section 4 integrates the aforementioned dimensions. Conclusions and future work end the paper in Sect. 5.

2 Identification of Relevant Dimensions

The CSHG describes and represents genetic knowledge with a rich level of detail. Nevertheless, to exploit it in a particular use case, such as representing the cardiomyopathies knowledge, we have to focus on the most relevant dimensions rather than the whole conceptual schema. To do it, we applied the ISGE method to obtain a conceptual schema that focuses on the relevant, identified dimensions. Figure 1 shows this schema, which we called *the Conceptual Schema of Cardiomyopathies* (CSC).

The classes that compose the CSC are grouped into the dimensions considered relevant. The **Location dimension** models any body location. It will be used for describing the heart's morphology. The LOCATIONS have been defined as a hierarchical composition that goes from the more general to the more specific. First, there are SYSTEMS, which cluster ORGANS that provide a specific functionality. Every ORGAN is composed of different types of TISSUES, which are made of millions of CELLS. CELLS, in turn, are constituted by biomolecules and structures called CELLULAR COMPONENTS.

The **Entity dimension** describes any biological entity of our body. It will be used to describe the elements that participate in the cardiac contraction process.

---

[1] Through grants INBIO 2021/AP2021-05 and INNEST/2021/57.

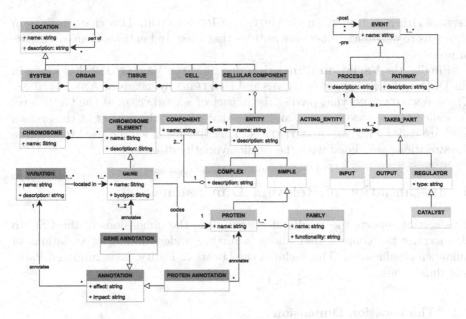

Fig. 1. Conceptual schema for cardiopathies obtained using the ISGE method. Red: Structural-related classes. Orange: Entity-related classes. Green: Location-related classes. Lilac: Pathway-related classes. (Color figure online)

A biological ENTITY is considered COMPLEX when it can be decomposed into smaller pieces that act as COMPONENTS. Otherwise, it is considered SIMPLE. There are different types of SIMPLE entities, but one of the most important is the PROTEINS because they play essential roles in the function, regulation, and structure of our body. Usually, PROTEINS are grouped into FAMILIES by functionality. Other examples of SIMPLE entities include ions and molecules such as ATP or ADP[2].

The **Structural dimension** describes the functional regions of our DNA. It will be used to describe the genes that codify the proteins that intervene in the heart's functionality and structure. Our DNA is arranged in CHROMOSOMES, that contain different CHROMOSOME ELEMENT with specific functionalities. One of these elements is the GENE, which is of high importance due to its function: codifying proteins.

The **Pathway dimension** describes the interactions of the biological ENTI-TIES. It will be used for describing the events that occur in the contraction of the heart. Compliant with the bio notation, these EVENTS can be either a PATH-WAY or a PROCESS, depending on whether they can be decomposed in simpler EVENTS. We defined non-rigid specializations [5] of the ENTITY, called ACTING ENTITY. These ACTING ENTITIES must TAKE PART in at least one specific PRO-

---

[2] Ions trigger some processes of the heart's muscle contraction while ATP and ADP are used by cells to obtain energy.

CESS as either an INPUT, an OUTPUT, or a REGULATOR. This characterization allows us to differentiate between entities that *exist* and entities that *participate* in processes.

Finally, the **Variation dimension** describes the VARIATIONS that occur in the DNA and their impact in GENES and PROTEINS by means of ANNOTATIONS. Those ANNOTATIONS that predict the impact of a VARIATION at the GENE level are called GENE ANNOTATION, and those that predict the impact at the protein level are called PROTEIN ANNOTATION. This dimension will be used to describe the variations associated with the cardiomyopathy disorder.

3   Instantiation of Relevant Dimensions

This section reports the results of instantiating the dimensions of the CSC to characterize the context that allows a further understanding of variations of unknown classification. This includes the Location, Entity, Structural and Pathway dimensions.

3.1   The Location Dimension

The heart is the organ responsible for pumping the blood through the body with rhythmic contractions and expansions. Figure 2 represents its structure at the morphological, cellular, and subcellular levels as an instantiation of the Location dimension of the CSC.

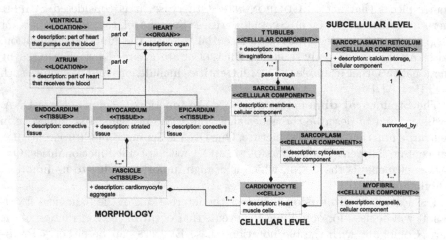

Fig. 2. Instantiation of the location dimension of the CSC.

At the **morphology level** the heart is divided into four chambers: two atria and two ventricles [6], and it is constituted by three different layers of tissue:

the *epicardium*, the *myocardium*, and the *endocardium*. We give special attention to the *myocardium*, the muscle tissue that is altered in patients with cardiomyopathies. The *myocardium* tissue is organized in *fascicles*, which are a bundle of cells enveloped together [11]. At the **cellular level**, the *fascicles* are constituted by specialized cardiac muscle cells called *cardiomyocytes*. Finally, at the **subcellular level**, the *cardiomyocytes* are constituted by a *sarcolemma* (cellular membrane), a *sarcoplasm* (cytoplasm) and a nucleus. Two important structures constitute the sarcoplasm: the *sarcoplasmatic reticulum*, responsible for providing the calcium that is needed calcium during the heart's contraction [6,11], and the *myofibrils*, which are the main actors in the myocardium contraction process.

3.2   The Entity Dimension

The myofibrils are constituted by a set of biological entities that, in this particular case, are protein complexes [6]. Figure 3 represents the entities that compose the myofibrils as an instantiation of the Entity dimension of the CSC.

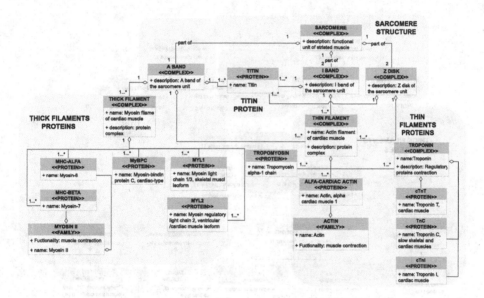

Fig. 3. Instantiation of the entity dimension of the CSC.

The protein complexes that constitute the myofibrils are called *sarcomeres*, and they are the basic unit for the myocardium contraction. Regarding the **sarcomere structure**, the proteins of the *sarcomere* are spatially arranged in three bands (one *A band*, and two *I band*), and are bounded together by two disks (*Z disks*) of proteic composition [6].

Regarding the *sarcomere* composition, its three bands are made up of a filament system composed of the **Thin filament** protein complex, the **Thick**

filament protein complex, and the **Titin** protein [7]. On the one hand, the *Thin filament* protein complex is mainly constituted by the *Alfa-Cardiac Actin* protein of the *Actin* family of proteins, and by other proteins such as the *Tropomyosin* protein, and the *Troponin* protein complex (constituted by the *cTnt*, *TnC*, and *cTnI* proteins) [9]. On the other hand, the *Thick filament* protein complex is mainly constituted by the *MHC-Alfa* and the *MHC-Beta* proteins of the *Myosin II* family, but other proteins such as the *MyBPC*, *MYL1*, and *MYL2* are also present [9]. Finally, the *Z disks* are mainly constituted by the Actinin, Capping protein, Desmin, and Enigma homolog proteins [7], but they are not represented here for simplicity.

3.3   The Structural Dimension

The proteins that constitute the sarcomere are codified by genes located in specific chrosomosomes. Figure 4 represents the genes that codify for the sarcomeric proteins and the chromosomes where these genes are located as an instantiation of the Structural dimension of the CSC.

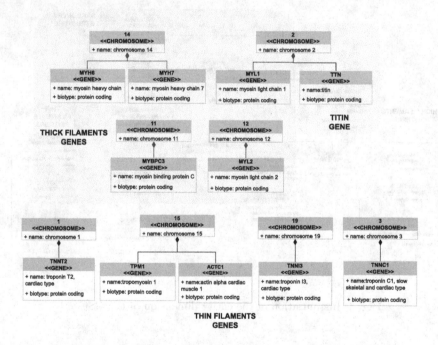

Fig. 4. Instantiation of the structural dimension.

The genes that code the proteins of the **Thick filament** are distributed in four chromosomes. The *chromosome 2* contains the *MYL1* gene, the *chromosome 12* contains the *MYL2* gene, the *chromosome 11* contains the *MYBC3* gene, and the *chromosome 14* contains the *MYH6* and *MYH7* genes.

The genes that code the proteins of the **Thin filament** are distributed in four chromosomes. The *chromosome 1* contains the *TNNT2* gene, the *chromosome 15* contains the *TPM1* and *ACTC1* genes, the *chromosome 19* contains the *TNNI3*, and *chromosome 3* the *TNNC1* gene.

Finally, the *chromosome 2* contains the *TTN* gene, that codes for the **Titin** protein.

3.4   The Pathway Dimension

The proteins that constitute the sarcomere are the main actors of the contraction process of the myocardium (i.e., *cardiac muscle contraction* pathway). As the cardiomyopathies affect the myocardium, its contraction process is often compromised. Figure 5 represents this contraction process as an instantiation of the Pathway dimension of the CSC.

The heart's contraction starts with the **binding of the Ca2+** ions to the Troponin complex of the Thin filaments of the sarcomere in the process *Ca2+ binds the troponin complex*. As a result, a *Ca2+ Sarcomere complex* is obtained.

After this, the **Myosin-Acting binding** stage begins. The union of the Ca2+ ions triggers the *Unblocking Actin union site* process, which consists of a change in the conformation of the Thin filaments that liberates several union sites where the Myosin II (either MHC-ALFA or MHC-BETA) in the Thick filaments can bind to the Alfa-Cardiac Actin protein of the Thin filaments. As a result, an *unblocked Ca2+ Sarcomere complex* is obtained. This stage ends with the binding of the Myosin protein to the Actin protein in the *Myosin binds Actin* process.

Finally, the **Sliding process** starts, and the Thick filaments slide through the Thin ones. To do this, first, ATP molecules bind to the Myosin of the Thick filaments in the *Myosin binds ATP* process, obtaining the *ATP - Sarcomere Complex* as an output. Second, these ATP molecules are hydrolyzed in the *ATP Hydrolisis by Myosin* process, which generates the an *ADP - Sarcomere Complex* and provides the energy necessary for the sliding process. Lastly, this energy is used in the *Myosin-Actin Binding and ADP release* process, where the ADP is released from the Myosin of the Thick filaments, and the Thin Filaments slide through the Thick ones. Then, the Myosin binds to the Alfa-Cardiac Actin in a new union site, and the sliding process begins again. As a result of the sliding process, the sarcomeres shorten, producing a muscle tension that is the basis for the myocardium contraction [8].

4   Integration of Relevant Dimensions

The CSHG describes how to integrate relevant dimensions to achieve data integration and semantic interoperability. Here, we illustrate the different connections established between the instantiated dimensions.

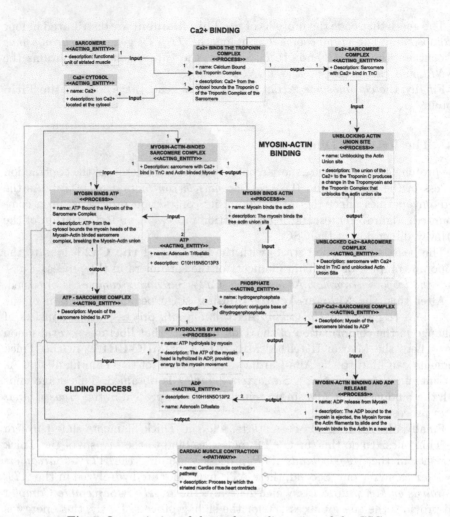

Fig. 5. Instantiation of the pathway dimension of the CSC.

Figure 6 depicts the four connections that exist between the dimensions, indicating which classes are connected and how they are connected (e.g., association, composition, or inheritance). The first connection represents that **variations are located in genes**. The variations identified in a patient are linked to the genes whose DNA sequence is altered. Therefore, the variation dimension and the structural dimension of the CSC are linked through the *located in* relationship between the Variation and the Gene class.

The second connection represents that **genes code for proteins**. A gene contains all the information needed to code for a specific protein. Consequently, the structural and the entity dimensions of the CSC are connected by the *codes* relationship between the gene and protein classes. For instance, the *MYH7* gene codes for the *MHC-BETA* protein (see Fig. 7a).

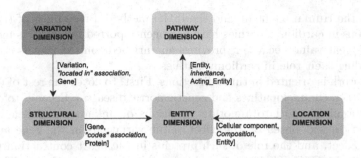

Fig. 6. Dimension connection.

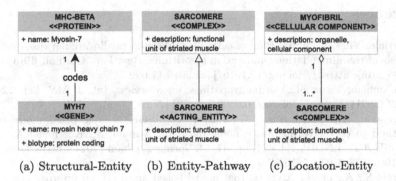

(a) Structural-Entity   (b) Entity-Pathway   (c) Location-Entity

Fig. 7. Examples of the dimension connection

The third connection represents that **biological entities take part in processes**. The entity and the pathway dimensions are connected by an *specialization* of the entities (i.e., acting entities) to represent their participation in a process. For instance, the *Sarcomere* protein complex specializes to a *Sarcomere* acting entity when describing its participation in the processes of the cardiac muscle contraction pathway (see Fig. 7b).

Finally, the last connection represents that **entities can be part of cellular components**. At the most elemental level, a cell is composed of cellular components, some of them being composed of biological entities such as proteins. We represent this connection between the entity and the location dimensions by an *aggregation* relationship between the Cellular component and Entity classes. For example, the *Sarcomere* protein complexes are part of the *Myofibril* cellular component (see Fig. 7c).

5 Conclusions and Future Work

This paper identified the conceptual dimensions of interest regarding cardiomyopathies, instantiated them to a practical case, and integrated them. A conceptual model-based approach has supported our efforts through the Conceptual

Schema of the Human Genome and the ISGE method. Since most of the studied variations in cardiomyopathies have not been reported before, this modeling effort is of great value because it provides an enriched context that is crucial in understanding their role in cardiomyopathies.

Future work is oriented in three directions. First, to cover the rest of the cardiopathies (i.e., channelopathies and genetic aortic diseases). Second, to develop a publicly available API to annotate variations of unknown significance with an enriched context based on the effect of the variations in the gene and proteins they affect, and the role of such proteins in the heart contraction process according to the CSC. Third, to include pharmacogenetics in the model.

References

1. Barriales-Villa, R., et al.: Plan of action for inherited cardiovascular diseases: synthesis of recommendations and action algorithms. Rev. Esp. Cardiol. **69**(3), 300–309 (2016). https://doi.org/10.1016/j.rec.2015.11.029
2. Ciarambino, T., et al.: Cardiomyopathies: an overview. Int. J. Mol. Sci. **22**(14), 7722 (2021). https://doi.org/10.3390/ijms22147722
3. García S., A., Casamayor, J.C., Pastor, O.: ISGE: a conceptual model-based method to correctly manage genome data. In: Nurcan, S., Korthaus, A. (eds.) CAiSE 2021. LNBIP, vol. 424, pp. 47–54. Springer, Cham (2021). https://doi.org/10.1007/978-3-030-79108-7_6
4. García S., A., et al.: A conceptual model-based approach to improve the representation and management of omics data in precision medicine. IEEE Access **9**, 154071–154085 (2021). https://doi.org/10.1109/ACCESS.2021.3128757
5. Guizzardi, G., Wagner, G., Almeida, J.P.A., Guizzardi, R.S.: Towards ontological foundations for conceptual modeling: the unified foundational ontology (UFO) story. Appl. Ontology **10**(3–4), 259–271 (2015)
6. Hall, J.E., Guyton, A.C.: Guyton and Hall Textbook of Medical Physiology. Saunders/Elsevier, Philadelphia (2011)
7. Henderson, C.A., et al.: Overview of the muscle cytoskeleton. Compr. Physiol. **7**(3), 891–944 (2017). https://doi.org/10.1002/cphy.c160033
8. Krans, J.L.: The sliding filament theory of muscle contraction. Nat. Educ. **3**(9), 66 (2010)
9. Morimoto, S.: Sarcomeric proteins and inherited cardiomyopathies. Cardiovasc. Res. **77**(4), 659–666 (2008). https://doi.org/10.1093/cvr/cvm084
10. Perry, E., et al.: Classification of the cardiomyopathies: a position statement from the European society of cardiology working group on myocardial and pericardial diseases. Eur. Heart J. **29**(2), 270–276 (2008). https://doi.org/10.1093/eurheartj/ehm585
11. van der Velden, J., Stienen, G.J.: Cardiac disorders and pathophysiology of sarcomeric proteins. Physiol. Rev. **99**(1), 381–426 (2019). https://doi.org/10.1152/physrev.00040.2017

A Conceptual Model of Health Monitoring Systems Centered on ADLs Performance in Older Adults

Francisco M. Garcia-Moreno[1] , Maria Bermudez-Edo[1(✉)] ,
José Manuel Pérez Mármol[2] , José Luis Garrido[1] ,
and María José Rodríguez-Fórtiz[1]

[1] Department of Software Engineering, Computer Sciences School, University of Granada,
C/Periodista Daniel Saucedo Aranda, s/n, 18014 Granada, Spain
{fmgarmor,mbe,jgarrido,mjfortiz}@ugr.es

[2] Department of Physiology, Faculty of Health Sciences, University of Granada, Av. de la
Ilustración, 60, 18016 Granada, Spain
josemapm@ugr.es

Abstract. Older adults usually present physical and mental problems such as anxiety, stress, depression, and mood disorders. In addition, there is a strong correlation between emotions/socialization and health. Negative emotions affect mental and physical health and can be caused by other diseases. Social isolation is a health risk factor comparable to smoking or physical inactivity. The diagnosis process is usually time-consuming and requires resources. The performance of the Activities of Daily Living (ADLs) could be used as an index of the decay of the elders, which can be delayed. ICTs can provide valuable and automatic support to health professionals facilitating routine tasks. Health monitoring systems, especially multi-sensing and intelligent, should be designed to fulfil requirements from each specific health domain. This paper reviews state-of-the-art and proposes a conceptual model centered on the ADLs concept, considering different health dimensions (social, emotional, physical and cognitive). Our proposal allows the evaluation of the elders' health holistically, and transparently. The conceptual model provides comprehensibility for this domain and provides a basis for developing multi-sensing and intelligent health monitoring systems.

Keywords: Health monitoring · Activities of Daily Living · Conceptual models

1 Introduction

The World Health Organization (WHO) defines the quality of life as "an individual's perception of their position in life in the context of the culture and value systems in which they live, and in relation to their goals, expectations, standards and concerns" [1]. For the older adults, good quality of life means living independently and feeling good while carrying out Activities of Daily Living (ADLs), considering physical, psychological, and social aspects of life [2]. Therefore, collecting information about physical, social,

R. Guizzardi and B. Neumayr (Eds.): ER 2022 Workshops, LNCS 13650, pp. 25–34, 2022.
https://doi.org/10.1007/978-3-031-22036-4_3

and cognitive areas is necessary to promote active and healthy ageing and provide people with an independent life for a longer time [3].

The early detection of behavioral changes and risk factors of functional decline can prevent problems in ageing. For example, detecting aspects restricting participating in the ADLs may be crucial to prevent reversible factors that different health agents such as physicians can approach. Depressive and anxiety symptoms are usually present in the elderly and cause a decrease in the quality of life in this population, favoring social isolation and the appearance of other clinical diseases. The ageing process associated with a sedentary lifestyle also favors mental, social, and physical issues because this may cause physiological changes such as muscle strength, aerobic capacity, and motor impairments. These aspects lead to a decrease in the capacity to perform ADLs efficiently and independently [2]. Diseases such as frailty and dependence can be predicted by observing when older adults reduce their performance level in ADLs [4]. Moreover, the ADL concept entails a holistic perspective because of the different dimensions (cognitive, social, emotional, etc.) required for performing these activities. Another interesting property of the monitoring of the performance of the ADLs is the ecological perspective. The term "ecological" refers to observe the person-in-the-environment (during their daily life) i.e., outside the laboratory or clinical setting. ADLs assessment interferes less with the daily life of older adults than traditional assessments, saves time, associated cost and is more efficient spending the health system resources.

Nowadays, there is an interest in political and research areas to improve the quality of life of elderly people. However, the previous observational studies and experiments are focused on isolated aspects of health, and the literature is limited in scope. Recent Internet of Things (IoT) solutions use mobile/wearable devices, platforms and systems, and data analytics. They are currently known as multi-sensing, intelligent systems. They collect data from sensors providing physiological information related to health (e.g. heart rate, skin temperature, movement, etc.) and from devices providing context information (e.g., presence in a room, geo-positioning, atmospheric pressure, etc.). These systems are ad hoc solutions for addressing some of the mentioned aspects or properties.

Conceptual Modelling has been recognized as an important method to manage complexity and, particularly in the fields of system analysis and design. Conceptual models have been used to provide formal (or semi-formal) representations of relevant aspects of the physical and digital realities. However, some works claim the need of reconceptualization of conceptual modelling in light of changing and emerging requirements nowadays [5]. One of these emerging requirements is to understand human needs and how design responds to these needs, i.e. human-centered design [6].

This paper reviews the state of the art and proposes a conceptual model centered on the ADL concept. Previous models focus on modelling the technological view (sensors, architectures, procedures, etc.) or on monitoring and recognizing specific ADLs, mainly indoors and involving only the physical dimension of health. The aim of this paper is to generalize the modelling of ADLs, considering the different dimensions (social, emotional, physical, and cognitive), required for performing the ADLs and to provide an integral solution. This conceptual model shapes the evaluation of the elderly's health holistically and ecologically, extending previous models found in the literature. Furthermore, this model deepens the knowledge of this specific domain and serves as a basis for developing intelligent and multi-sensing health monitoring systems.

The paper is organized as follows. Section 2 presents foundations focused on the important aspects of elderly people's health monitoring and ADLs. Section 3 reviews related work. Section 4 presents the proposed conceptual model and instantiation example. Finally, Sect. 5 summarizes conclusions and introduces future research.

2 Foundations

The term ADLs involves all activities performed by human beings during their lifespan. One of the most accepted classification for ADLs is based on their level of complexity. This classification organizes activities from the basic (BADL) (e.g. self-care activities, functional mobility, and the care of personal devices), through the instrumental (IADL) (e.g. use of the phone, shopping tasks, and use of transportation), and up to the advanced (AADL) (e.g. planning travels, and participation in events and meetings). Each ADL requires different body functions and structures from different dimensions (physical, cognitive, social and emotional). IADLs require cognitive and motor complexity (executive functions), and imply an interaction with the social environment that surrounds the persons [7]. AADLs are the most complex ADLs as they involve voluntary physical and social functions, but are not essential to maintain independence. The performance of IADLs is an important health indicator to predict mild or severe cognitive impairments, such as dementia in older adults [8].

Health systems can evaluate the health status by observing people's movement and exercise intensity, recognizing indoor and outdoor ADLs, and even detecting food intake, interactions with relatives and friends, etc. The identification of risks and anomalies is important to help the elderly and caregivers to prevent dangerous situations [3]. Health systems based on intelligent systems aim to reduce hospitalized demands and costs. They include sensors that allow the continuous monitoring of different aspects of health such as the vital parameters, physical activities, and falls [9].

IoT includes sensors and devices with sensors, such as wearables and smartphones, used in the health environments to monitor biological, behavioral, and environmental data of people, because they are non-invasive, easily acceptable by subjects, and do not intrude users in their normal activities [10]. The most common sensors used in health are [9]: electrodermal activity (EDA), photoplethysmography (PPG), electrocardiography (ECG), electroencephalography (EEG), and skin temperature (SKT), among others. They collect physiological data such as heart rate, blood pressure, body temperature, respiratory rate, and blood oxygen saturation. Other physical reactions can also be measured with video or infrared cameras, microphones and electromyograms (EMG), for example, facial and body gestures. They can be combined with technologies to identify the dynamic position of people: Radio-Frequency Identification (RFID), Bluetooth Low Energy (BLE) or beacons, GPS, accelerometer (ACC) and gyroscope (GYR). Environmental sensors and sensors embedded in furniture, home appliances, walls and carpets also help to gather contextual information useful to know the user's behavior. Once the sensors collect the data, it is necessary to analyze it to infer relevant information.

Machine learning (ML) techniques are common in these situations because they provide better accuracy with big quantities of sensory data than other statistical analyses [11]. In particular, ML has been used to accurately recognize activities, detect health risk factors and specific health conditions such as frailty or dependence [12, 13].

There are some important challenges in the development of the health monitoring systems: usability improvement, low cost-based solutions, data security guarantees, integration of devices, quality data collecting and processing, managing big data, and device power consumption [14]. Conceptual modelling could be useful to analyze systems complexity and also to estimate the strategy behind the development of the software and the best devices to be used. Some of those challenges can be addressed by including them in the model.

3 Related Work

Several IoT initiatives tried to model the sensor environments, and even some standardization bodies have proposed their solutions [15]. It is a field in continuous evolution and the reuse, evolution and extension of previous models is a common practice. One of the main initiatives is the Semantic Sensor Network model (SSN) by W3C, which gathered all sensor models at that time and developed a complete semantic model centered on the concept of the sensor [16]. It also includes other concepts and relationships between them, such as sensor properties, systems, deployments, stimuli, and observations. Other IoT models evolve by extending SSN, such as IoT-A [17], which introduces the concept of service, resources, and entities, or SAO [18], for data analytics and event detection.

However, with the proliferation of sensors and stream data in IoT, SSN was too heavy to effectively process IoT data. Then, other initiatives move to the lightweight models. For example, IoT-Lite [19], which also introduced the concepts of actuators and coverage. LiO-IoT [20] that extends IoT-Lite adding Tag concepts and relationships. IoT-Stream [21], which deals with analytics. Even W3C updated its SSN with a core model less heavy, Sensor, Observation, Sample and Actuation (SOSA) [22].

Other models complement the sensor models with concepts needed to annotate the sensors and specifically the sensory data. For example, location models, such as Geo[1] locate the sensors or data. Geo is composed of a few basic terms, such as latitude, longitude, and altitude. GeoSPARQL is a standard for the representation and querying of geospatial data from the Open Geospatial Consortium (OGC) [23]. GeoJSON[2] is another example that describes geographic features, and geometric forms, such as Point, LineString, Polygon, MultiPoint, etc. Time ontology[3] represents topological (ordering) relations, duration, and temporal position (i.e., date–time information) with different time references (unix, geologic time, etc.).

All these models need taxonomies, which categorize the different devices, activities, etc. For example, some taxonomies of sensors describe the characteristics of sensors: power, configuration, material, sensing methods, functions, etc. [24]. The QU[4] model

---

[1] https://www.w3.org/2003/01/geo/.
[2] https://tools.ietf.org/html/rfc7946.
[3] https://www.w3.org/TR/2017/REC-owl-time-20171019.
[4] https://www.w3.org/2005/Incubator/ssn/ssnx/qu/qu-rec20.html.

focuses on quantities and units and supports different Systems Modelling Languages (SysML) users. Even some initiatives perform a step further and do not only create the taxonomy, but divide the taxonomy in levels. For example, [25] presents a wristband taxonomy in three levels: the tracking raw input (sensory data), the raw output (devices and HW), and the intelligent output (high-level events).

Dealing with sensory data is usually noisy and faulty, so we need to model the Quality of Information as well. The common quality concepts modelled are Completeness, Correctness, Concordance, Currency, Plausibility [26] and Security [27], access control in cloud data [28] and the provenance of the data, PROV-O [29]. Even Zero model [30] uses blockchain to assure the security and traceability of data.

In the field of health, Health Level-7 (HL7) regulates the digital transfer of clinical and administrative data. It is a set of international standards and guidelines which provides a common vocabulary in order to interoperate between the endpoints. One of these standards is the Fast Healthcare Interoperability Resources (FHIR) (https://www.hl7.org/fhir/index.html), which divides the concepts into seven levels. Level 1 provides the basic framework, mainly data types and formats. Level 2 deals with implementation and binding to external specifications, with concepts such as versions, databases, security. Level 3 defines the patients and other concepts of healthcare systems such as devices, locations, etc. Level 4 annotates the records and processes, such as diagnosis, medication and financial issues. And level 5 provides the reasoning, modelling concepts such as actuation plans.

Several initiatives have already merged the concepts mentioned above [31]. For example, SmartEnv has applied them to older adults, extending SSN and representing different aspects of smart and sensorized environments [32]. These aspects are: observation/sensing, agents, activities/events, objects, network set-up, spatial and temporal aspects. The model annotates autonomous health systems used for elderly homes.

Regarding ADLs, [33] models regular activities performed in a house, with actions and concepts. For example, calling someone, etc. But it only models some activities and no instrumental activities. [34] models the human activities in smart homes, emphasizing the time sequence, i.e. whether the activities are sequential or concurrent. Additionally, they use rules to detect activities, for example sleep takes place in bed and has a duration of 8 h. However, these models only focus on BADLs performed in the house.

4 A Conceptual Model Centered on ADLs Performance

We propose a holistic and ecological model for health focused on the monitoring of ADLs performance with sensors and data analytics. This model extends and reuses the models presented in Sect. 3, borrowing some concepts and relationships. Figure 1 shows the general overview of the model, and Fig. 2, the important concepts, and relationships.

Our model has 3 layers: hardware, software, and domain.

The **hardware layer** includes devices (e.g., wearables, mobile phones) and sensors used to collect data and to show information to the users. Table 1 shows one column for each concept: devices, sensors, and characteristics. A device can have several sensors embedded or receive data from external sensors. We have represented only the sensors most used in health: physiological and environmental sensors.

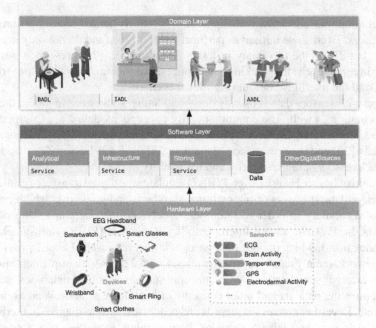

Fig. 1. General overview of the conceptual model

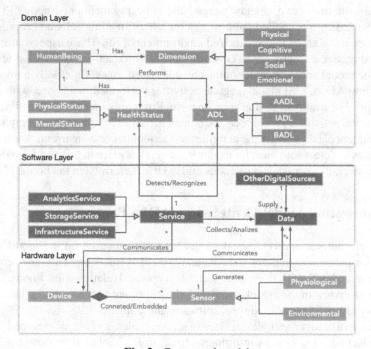

Fig. 2. Conceptual model

Table 1. Characteristics of core concepts of the hardware layer

Device	Sensor	Measuring capabilities
Name	Name	Coverage
OperatingSystem	Domain	Latency
NetworkConnectivity	Configuration	Accuracy
Screen	Power	Frequency
Battery	SensingFunction	Resolution
Memory	SensingMethod	Sensitivity
ComputingPower	Material	Selectivity
	ObservedProperty	Precision
	StimulusDetected	ResponseTime
	Type	DetectionLimit
		inCondition

The **software layer** consists of data that are generated by sensors, communicated by devices or supplied by other digital resources such as external services (See Table 2). Each data has a timestamp, type, rank, value, and several quality attributes. This layer also includes a set of services that manage data. There are specific services to communicate data between devices and servers, in which the services are deployed. Most common communication services implement request/reply (point-to-point communication) or Publication/Subscription (many-to-many communication) paradigms. Other specific services are specialized in data storing, for instance, data could be stored in a non-SQL database (local or cloud). And other analytic services process the data which usually apply machine learning techniques to identify or classify health status. The results of the analysis are sent to the devices used by elders or caregivers.

Table 2. Characteristics of core concepts of the software and domain layer

Service	Data	ADL	Health state
Name	Timestamp	Name	Factor
Type/method/technique	Type	Type	Type
Parameters []	Value	Description	Rank
Performance indexes []	Rank	Place	Value
	Format	Time sequence	Health status
	Quality attribute []	Measurements []	Factor

The **domain layer** represents the human beings and the information of their health status and ADLs performance. There are three types of ADLS: AADLs, IADLs and

BADLs. Each ADL has a name, a description and is performed in a place or a space (indoor or outdoor). Besides, an ADL can be carried out at the same time as others or sequentially (time sequence) and has different aspects that can be measured to assess its performance (Motions, Gestures, BodyAcceleration, Proximity, Duration, Intensity, etc.). We differentiate between two types of physical and mental health status, with different values for each health factor. Besides, human beings can be observed from different dimensions, (physical, cognitive, social and emotional). Table 2 also shows the main characteristics for the core concepts of this layer, ADL and Health state.

5   Conclusions

The decay in the elderly, which can be delayed or reversed, affects not only the physical dimension of health but also the cognitive, emotional, and social dimensions. We could consider all these dimensions by monitoring the elderly during their ADLs. Additionally, this monitoring is holistic and ecological. Systems using sensors and data analytics can help in the automation of the health assessment by monitoring the performance of the ADLs.

To the best of our knowledge, none of the previous models provides all the concepts and relationships we need, and none centers on the IADLs performed by the elderly indoors and outdoors. We have used several concepts and relationships from the previous literature in our solution. We have added the missing parts and linked them together. Our model extends the literature by completing a comprehensive model to monitor health by evaluating IADLs performance with their dimensions (physical, cognitive, etc.). The model provides a generic and integral solution and support ecologic and holistic health evaluations.

Hence, a holistic and ecological evaluation system for monitoring elderly people can be established by using observations of the elders in their natural contexts during the performance of the ADLs involved in their routines. This novel monitoring system may serve as a form for the evaluation of multiple health's aspects at the same time, reducing time and cost for the health system and professionals. Additionally, the behavior of elderly people under evaluation is expected to be more accurate/precise about the real health status due to the reduction of multiple possible biases happening during the evaluation of the person in a clinical setting. The conceptual model presented in the current work covers all the relevant elements to develop this monitoring system. Our model could serve as a basis for developing ecological and holistic health monitoring systems that consider several dimensions of health. These systems help health professionals to automate the assessment of the health status. In the future we want to validate the proposed model implementing a monitoring system and performing an extensive experiment with elderly people.

Acknowledgements. This research is funded by: Junta of Andalucia and European Regional Development Funds (FEDER, UE) through the project B-TIC-320-UGR20, and by the Spanish Ministry of Science and Innovation through the project Ref. PID2019-109644RB-I00/AEI/10.13039/501100011033.

References

1. World Health Organization: Health statistics and information systems. WHOQOL: Measuring Quality of Life (2020). https://www.who.int/tools/whoqol
2. de Oliveira, L.D.S.S.C.B., Souza, E.C., Rodrigues, R.A.S., Fett, C.A., Piva, A.B.: The effects of physical activity on anxiety, depression, and quality of life in elderly people living in the community. Trends Psychiatry Psychother. **41**, 36–42 (2019)
3. Cicirelli, G., Marani, R., Petitti, A., Milella, A., D'Orazio, T.: Ambient assisted living: a review of technologies, methodologies and future perspectives for healthy aging of population. Sensors **21**(10), 3549 (2021)
4. De Vriendt, P., Gorus, E., Cornelis, E., Velghe, A., Petrovic, M., Mets, T.: The process of decline in advanced activities of daily living: a qualitative explorative study in mild cognitive impairment. Int. Psychogeriatr. **24**, 974–986 (2012)
5. Recker, J.C., Lukyanenko, R., Jabbari Sabegh, M., Samuel, B., Castellanos, A.: From representation to mediation: a new agenda for conceptual modeling research in a digital world. MIS Q.: Manag. Inf. Syst. **45**(1), 269–300 (2021)
6. Zhang, T., Dong, H.: Human-centered design: an emergent conceptual model. In: Include2009 Proceedings. Include2009, Royal College of Art, April 8–10, 2009, London (2009). ISBN: 978-1-905000-80-7
7. Lawton, M.P., Brody, E.M.: Assessment of older people: self-maintaining and instrumental activities of daily living. Gerontol. **9**(3_Part_1), 179–186 (1969)
8. Pashmdarfard, M., Azad, A.: Assessment tools to evaluate Activities of Daily Living (ADL) and Instrumental Activities of Daily Living (IADL) in older adults: a systematic review. Med. J. Islam Repub. Iran **34**, 33 (2020)
9. Smuck, M., Odonkor, C.A., Wilt, J.K., Schmidt, N., Swiernik, M.A.: The emerging clinical role of wearables: factors for successful implementation in healthcare. NPJ Digit. Med. **4**(1), 1–8 (2021)
10. Dzedzickis, A., et al.: Human emotion recognition: review of sensors and methods. Sensors **20**(3), 592 (2020)
11. Mahdavinejad, M.S., Rezvan, M., Barekatain, M., Adibi, P., Barnaghi, P., Sheth, A.P.: Machine learning for Internet of Things data analysis: a survey. Digit. Commun. Netw. **4**(3), 161–175 (2018)
12. Shinde, S.A., Rajeswari, P.R.: Intelligent health risk prediction systems using machine learning: a review. Int. J. Eng. Technol. **7**(3), 1019–1023 (2018)
13. Garcia-Moreno, F.M., Bermudez-Edo, M., Rodríguez-García, E., Pérez-Mármol, J.M., Garrido, J.L., Rodríguez-Fórtiz, M.J.: A machine learning approach for semi-automatic assessment of IADL dependence in older adults with wearable sensors. Int. J. Med. Inform. **157**, 104625 (2022)
14. AlShorman, O., Alshorman, B., Masadeh, M., Alkahtani, F., Al-Absi, B.: A review of remote health monitoring based on internet of things. Indonesian J. Electr. Eng. Comput. Sci. **22**(1), 297–306 (2021)
15. Zorgati, H., Djemaa, R.B., Amor, I.A.B., Sedes, F.: QoC enhanced semantic IoT model. In: Proceedings of the 24th Symposium on International Database Engineering & Applications, pp. 1–7 (August 2020)
16. Compton, M., et al.: The SSN ontology of the W3C semantic sensor network incubator group. J. Web Semant. **17**, 25–32 (2012)
17. Bassi, A., et al.: Enabling Things to Talk: Designing IoT Solutions with the IoT Architectural Reference Model. Springer, Heidelberg (2013). https://doi.org/10.1007/978-3-642-40403-0
18. Kolozali, S., Bermudez-Edo, M., Puschmann, D., Ganz, F., Barnaghi, P.: A knowledge-based approach for real-time IoT data stream annotation and processing. In: Proceedings of the 2014

IEEE International Conference on Internet of Things (iThings), and Green Computing and Communications (GreenCom), IEEE and Cyber, Physical and Social Computing (CPSCom), Taipei, Taiwan, 1–3 September 2014, pp. 215–222 (2014)

19. Bermudez-Edo, M., Elsaleh, T., Barnaghi, P., Taylor, K.: IoT-Lite: a lightweight semantic model for the internet of things and its use with dynamic semantics. Pers. Ubiquit. Comput. **21**(3), 475–487 (2017). https://doi.org/10.1007/s00779-017-1010-8

20. Rahman, H., Hussain, M.I.: A light-weight dynamic ontology for Internet of Things using machine learning technique. ICT Express **7**(3), 355–360 (2021)

21. Elsaleh, T., Enshaeifar, S., Rezvani, R., Acton, S.T., Janeiko, V., Bermudez-Edo, M.: IoT-Stream: a lightweight ontology for Internet of Things data streams and its use with data analytics and event detection services. Sensors **20**(4), 953 (2020)

22. Janowicz, K., Haller, A., Cox, S.J., Le Phuoc, D., Lefrançois, M.: SOSA: a lightweight ontology for sensors, observations, samples, and actuators. J. Web Semant. **56**, 1–10 (2018)

23. Battle, R., Kolas, D.: Enabling the geospatial semantic web with parliament and geosparql. Semant. Web **3**, 355–370 (2012)

24. Haddara, Y.M., Howlader, M.M.: Integration of heterogeneous materials for wearable sensors. Polymers **10**(1), 60 (2018)

25. Kamišalić, A., Fister, I., Jr., Turkanović, M., Karakatič, S.: Sensors and functionalities of non-invasive wrist-wearable devices: a review. Sensors **18**(6), 1714 (2018)

26. Weiskopf, N.G., Weng, C.: Methods and dimensions of electronic health record data quality assessment: enabling reuse for clinical research. J. Am. Med. Inform. Assoc. **20**, 144–151 (2013)

27. Gonzalez-Gil, P., Skarmeta, A.F., Martinez, J.A.: Towards an ontology for IoT context-based security evaluation. In: Proceedings of the 2019 Global IoT Summit (GIoTS), Aarhus, Denmark, 17–21 June 2019, pp. 1–6 (2019)

28. Dutta, S., Chukkapalli, S.S.L., Sulgekar, M., Krithivasan, S., Das, P.K., Joshi, A.: Context sensitive access control in smart home environments. In: 2020 IEEE 6th International Conference on Big Data Security on Cloud (BigDataSecurity), IEEE International Conference on High Performance and Smart Computing (HPSC) and IEEE International Conference on Intelligent Data and Security (IDS), pp. 35–41. IEEE (May 2020)

29. Lebo, T., et al.: PROV-O: The PROV Ontology. W3C Recommendation, W3C (2013). https://www.w3.org/TR/prov-o/. Accessed June 2022

30. Sandhiya, R., Ramakrishna, S.: Investigating the applicability of blockchain technology and ontology in plastics recycling by the adoption of ZERO plastic model. Mater. Circ. Econ. **2**(1), 1–12 (2020)

31. Esnaola-Gonzalez, I., Bermúdez, J., Fernandez, I., Arnaiz, A.: Ontologies for observations and actuations in buildings: a survey. Semant. Web **11**(4), 593–621 (2020)

32. Alirezaie, M., Hammar, K., Blomqvist, E., Nyström, M., Ivanova, V.: SmartEnv ontology in E-care@ home. In: Proceedings of the 9th International Semantic Sensor Networks Workshop, Monterey, CA, USA, 9 October 2018, pp. 72–79 (2018)

33. Bae, I.H.: An ontology-based approach to ADL recognition in smart homes. Futur. Gener. Comput. Syst. **33**, 32–41 (2014)

34. Ni, Q., Pau de la Cruz, I., Garcia Hernando, A.B.: A foundational ontology-based model for human activity representation in smart homes. J. Ambient Intell. Smart Environ. **8**(1), 47–61 (2016)

A Comparative Analysis
of the Completeness and Concordance
# of Data Sources with Cancer-Associated
Information

Mireia Costa[✉] , Alberto García S. , and Oscar Pastor

PROS Group, Valencian Research Institute (VRAIN), Universitat Politècnica de
València, Camí de Vera, s/n, 46022 València, Spain
micossan@vrain.upv.es, {algarsi3,opastor}@pros.upv.es

Abstract. Precision medicine promises to improve the diagnosis and
treatment of patients based on their genetic particularities. One of the
fields in which clinicians aim to use precision medicine is oncology. The
knowledge that is necessary to achieve a proper precision oncology application is dispersed over many data sources with heterogeneous data representations. In this work, we studied seven of the most relevant data
sources used to deliver precision oncology and determined if the information contained in them is complete and concordant. The results herein
reported indicate that this information is neither complete nor concordant. Thus, providing proper precision oncology is still an unresolved
challenge for clinicians.

Keywords: Precision oncology · Comparative analysis · DNA
variations

1 Introduction

Precision medicine has emerged as a promising approach to improving patients'
diagnosis and treatment in all medical specialties, including oncology. The application of precision medicine in oncology (i.e., precision oncology) is crucial to
provide patients with personalized diagnoses and treatments based on their
genetic information. However, precision oncology is particularly challenging from
a genetic perspective because it not only deals with the study of germline DNA
variations (i.e., those that are inherited) but also with somatic variations (i.e.,
those that appear during the patient's life and are often found in tumor tissues).

The knowledge that a proper precision oncology application requires about
cancer-related variations is dispersed over several data sources [5]. For the purpose of this work, we classified these data sources into two types, **generic data
sources** and **cancer-specific sources**. Generic data sources contain information about variations associated with any disorder. Cancer-specific sources focus

Supported by ACIF/2021/117, MICIN/AEI/10.13039/501100011033, and INNEST/
2021/57 grants.

R. Guizzardi and B. Neumayr (Eds.): ER 2022 Workshops, LNCS 13650, pp. 35–44, 2022.
https://doi.org/10.1007/978-3-031-22036-4_4

solely on variations that influence cancer development, progression, or treatment. Each data source stores information using different structures and data formats. Consequently, differences in interpreting the clinical importance of cancer-related variations can emerge depending on the data sources consulted [14].

These differences complicate applying precision oncology. To further analyze this situation, we have conducted a comparative analysis of seven data sources (four generic and three cancer-specific) containing information about cancer-related variations. Specifically, we have evaluated the **completeness** of each data source's information and the degree of **concordance** among this information. Our study has been performed in the context of three cancer disorders (i.e., *Oligoastrocytoma*, *Oligodendroglioma*, and *Pleomorphic Xanthoastrocytoma*) that we are currently studying in a real-world use case together with oncologists from different spanish hospitals[1].

Even though there is literature describing comparative analyses of data sources containing information about cancer-related variations, they only compare what we have called cancer-specific sources [10, 11]. Thus, the novelty of our work is that we consider both generic and cancer-specific data sources, which has allowed us to deliver a more complete and richer analysis.

The remainder of the work is structured as follows: Sect. 2 describes the data sources selected for the analysis and the methodology followed. Section 3 presents the results of our analysis. Section 4 discusses the results. Finally, Sect. 5 presents conclusions and future work.

2   Materials and Methods

To conduct our comparative analysis, we selected four generic and three cancer-specific data sources and studied the differences in the information they provide for the three selected disorders.

The generic data sources are ClinVar, Ensembl, LOVD, and GWAS Catalog. ClinVar is a public archive that contains information about the role of germline and somatic variations in clinical health [8]. Ensembl is an infrastructure that aggregates information from different sources about the vertebrates genome, including the information stored in ClinVar [3]. LOVD is a source that clusters several gene-specific databases that contain germline and somatic variation-disease clinical relationships [4]. GWAS Catalog follows a different approach by storing GWAS studies data (i.e., statistical associations between a variation and a particular phenotype in a given population) related to germline variations [1].

The cancer-specific data sources are COSMIC, CIViC, and OncoKB. COSMIC is the most comprehensive resource for obtaining data about the role of somatic variations in cancer development [15]. CIViC is an open knowledgebase that provides interpretations about the relevance of germline and somatic variations for cancer disorders at different clinical relevant dimensions [6]. OncoKB is a curated database that provides evidence-based information about the role of somatic variations for cancer disorders at different clinical dimensions [2].

---

[1] Through grant INNEST/2021/57.

The comparative analysis evaluated the completeness and concordance of the selected sources. Completeness is defined as *the extent to which data are in sufficient breadth, depth, and scope for the task at hand* [16]. Concordance is defined as *the degree of agreement or compatibility between data elements* [17].

Completeness was studied at the schema and the data level [13]. **Completeness at the schema level** evaluates if the internal schema of the data source allows for representing all of the clinical dimensions that cancer-related variations have. Completeness at the schema level is disorder-independent as it depends on the internal data source schemes. **Completeness at the data level** evaluates whether a data source contains all of the cancer-related variations reported in the other studied data sources for a particular disorder and clinical dimension.

We adopted the clinical dimensions described in [9] (i.e., diagnostic, prognostic, and predictive). The **diagnostic** dimension considers information regarding the role of variations in causing a specific type of cancer. The **prognostic** dimension provides information for the role of variations in cancer progression or survival. The **predictive** dimension describes information for the role of variations in the response to a specific drug or treatment.

We evaluated completeness at the schema level by studying whether a data source allows representing information about cancer-related variations in the three dimensions mentioned above. Completeness at the data level was evaluated by comparing the variations that each data source provides in each of the three dimensions for each of the studied disorders.

Finally, we evaluated concordance among the data sources' information for each of the selected disorders in the three dimensions mentioned above. For each disorder and clinical dimension, we compared those variations that appear in more than one data source. Concordance is achieved if the information of such variations among the data sources is entirely coincident.

3   Results

3.1   Completeness at the Schema Level

Table 1. Completeness at the schema level.

Source	Diagnostic	Prognostic	Predictive	Completeness (CSL)
ClinVar	✓	X	✓	66,67%
Ensembl	✓	X	✓	66,67%
GWAS Catalog	✓	X	✓	66,67%
LOVD	✓	X	X	33,33%
COSMIC	✓	X	✓	66,67%
CIViC	✓	✓	✓	100%
OncoKB	✓	✓	✓	100%

Table 1 shows each data source's completeness at the schema level (CSL). Every generic data source contains information regarding the diagnostic dimension. Almost all of the generic data sources have information about the predictive dimension. The exception is LOVD, which only reports on the diagnostic dimension (CSL of 33,33%). Regarding cancer-specific data sources, CIViC and OncoKB provide information for the three dimensions (CSL of 100%), and COSMIC contains information about the diagnostic and predictive dimensions (CSL of 66,67%). All of the data sources except the GWAS Catalog represent the role of a variation in each of these dimensions with the concept of *clinical significance*. This concept is represented as an attribute whose value represents the role of a variation in a particular disorder. In GWAS Catalog, a *statistical association* is provided instead of the clinical significance.

In the diagnostic dimension, the terminology used for the clinical significance is not homogeneous across the data sources. For instance, a variation that causes a cancer disorder is represented using: the *pathogenic* value in ClinVar, Ensembl, LOVD, and COSMIC; the *Positive* value in CIViC; and a *text-free* description of the precise effect of the variation in OncoKB.

OncoKB and CIViC are the data sources providing clinical significance for the prognostic dimension. OncoKB uses a *text-free* description, whereas CIViC allows for two values: *better outcome* or *poor outcome*.

For the predictive dimension, the representation of the clinical significance is even more heterogeneous than in the others. ClinVar and Ensembl provide the *drug response* value. The problem with this value is that it does not reflect whether a variation is responsible for an enhanced or an adverse response to the treatment. COSMIC provides information for variations that affect 15 specific genes with the *drug resistance* value. CIViC offers the most comprehensive representation of the clinical significance of the predictive dimension, providing a set of values (i.e., *Sensitivity/Response*, *Resistance*, *Adverse response*, or *Reduced Sensitivity*) that allows for a precise assessment of the effect of a variation in treatment response.

3.2 Completeness at the Data Level

To study the **completeness at the data level (CDL)**, we compared the variations of each data source for the three studied disorders. The number of variations available in each source for each disorder is represented in Fig. 1.

Figure 1 reflects critical differences between the information in each data source. Regarding the generic data sources, ClinVar provides the most variations, with information about the Pleomorphic Xanthoastrocytoma and Oligodendroglioma disorders. Although Ensembl obtains its information from ClinVar, it only contains information for Oligodendroglioma, probably because Ensembl updates less frequently. GWAS Catalog and LOVD are not considered in the CDL assessment because they do not have information about any of the three analyzed disorders. Cancer-specific data sources provide a significantly higher number of variations. COSMIC provides the most variations for the Oligodendroglioma and Oligoastrocytoma disorders. For the Pleomorphic Xanthoastrocytoma disorder, OncoKB provides more variations. Next, we analyze the information on a per disorder basis.

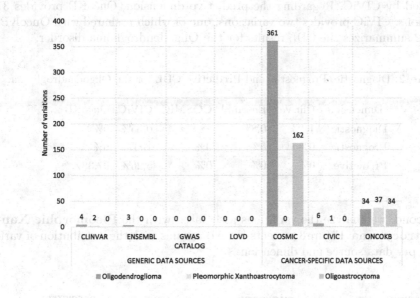

Fig. 1. Number of variations per database and disorder.

First, we analyze the variations associated with the **Oligodendroglioma** disorder. Figure 2 presents Venn diagrams with the distribution of variations per data source and dimension.

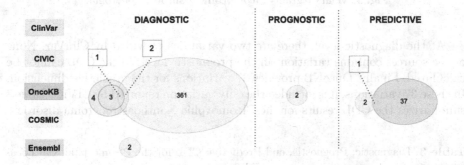

Fig. 2. Venn diagrams Oligodendroligoma.

In this disorder, most information concentrates on the diagnostic dimension, where COSMIC is the data source with the most variations (i.e., 361 variations). From these variations, one appears in both ClinVar and Ensembl, and another one appears in ClinVar alone. As expected, all of the variations in Ensembl are in ClinVar too. CIViC contains two variations that are not available in any other data source. For the prognostic dimension, only two variations are available, both provided by CIViC. Regarding the predictive dimension, OncoKB provides 37 variations. CIViC provides two variations, one of which is shared with OncoKB. Table 2 summarizes the CDL results for the Oligodendroglioma disorder.

Table 2. Diagnostic, Prognostic, and Predictive CDL for the Oligodendroglioma.

Dimension	ClinVar	Ensembl	COSMIC	CIViC	OncoKB
Diagnostic	1,1%	0,8%	98,9%	0,55%	0%
Prognostic	0%	0%	0%	100%	0%
Predictive	0%	0%	0%	5,26%	97,38%

Second, we analyze the variations associated with the **Pleomorphic Xanthoastrocytoma**. Figure 3 presents Venn diagrams with the distribution of variations per data source and dimension.

Fig. 3. Venn diagrams Pleomorphic Xanthoastrocytoma.

At the diagnostic level, there are two variations provided by ClinVar. None of the sources contains variations in the prognostic level, so the CDL cannot be calculated. Finally, OncoKB provides 37 variations for the predictive dimension. In these 37 variations, it is included the only variation reported by CIViC. Table 3 summarizes the CDL results for the Pleomorphic Xanthoastrocytoma disorder.

Table 3. Diagnostic, Prognostic, and Predictive CDL for the Pleomorphic Xanthoastrocytoma.

Dimension	ClinVar	Ensembl	COSMIC	CIViC	OncoKB
Diagnostic	100%	0%	0%	0%	0%
Prognostic	–	–	–	–	–
Predictive	0%	0%	0%	2,7%	100%

Finally, we analyze the variations associated with the **Oligoastrocytoma** disorder. Figure 4 presents the Venn with the distribution of variations per data source and dimension.

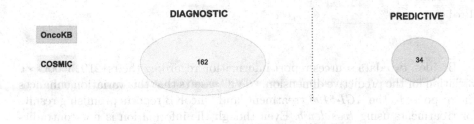

Fig. 4. Venn diagrams Oligoastrocytoma.

Two data sources provide variations related to the Oligoastrocytoma. The first data source, COSMIC, contains all of the variations associated with the diagnostic dimension (i.e., 162 variations). The second data source, OncoKB, provides all the variations associated with the predictive dimension (i.e., 34 variations). There are no variations associated with the prognostic dimension. Table 4 summarizes the CDL results for the Oligoastrocytoma disorder.

Table 4. Diagnostic, Prognostic, and Predictive CDL for the Oligoastrocytoma.

Dimension	ClinVar	Ensembl	COSMIC	CIViC	OncoKB
Diagnostic	0%	0%	100%	0%	0%
Prognostic	–	–	–	–	–
Predictive	0%	0%	0%	0%	100%

3.3 Concordance

To evaluate **concordance**, we compared the information associated with the variations available in more than one data source.

There are four coincidences in the information associated with the diagnostic dimension for the **Oligodendroglioma**. Three of these variations (i.e., rs1569535987T>C, rs1568503055insT, and rs121913500C>T) appear in both ClinVar and Ensembl, and their information is entirely concordant. Two variations appear in ClinVar and COSMIC (i.e., rs121913500C>T, and rs1568504941-C>T), being the former of these variations also available in Ensembl. Regarding the rs121913500C>T variation, ClinVar and Ensembl provide the *not provided* clinical significance. However, COSMIC reports it as *pathogenic*. Regarding rs1568504941C>T variation, both ClinVar and COSMIC provide a *pathogenic* clinical significance.

Table 5. Diagnostic, Prognostic, and Predictive concordance

Dimension	Oligodendroglioma	Pleomorphic Xanthoastrocytoma	Oligoastrocytoma
Diagnostic	75%	–	–
Prognostic	–	–	–
Predictive	0%	0%	–

Besides, two data sources report information regarding the rs121913500C>T variation for the predictive dimension. CIViC asserts that this variation enhances the response to the *AGI-5198* treatment, and OncoKB reports promising results for treatments using *Ivosidenib*. Even though the information is not contradictory, it is not coincident. The concordance cannot be assessed for the prognostic dimension as there are no common variations in the data sources.

In the **Pleomorphic Xanthoastrocytoma**, there is one variation (i.e., rs113488022T>A) that appears in more than one data source (see Fig. 3), and it is associated with the predictive dimension. This variation is reported as producing an enhanced response to the *Vemurafenib* treatment in CIViC. However, OncoKB indicates that this enhanced response only appears when combining the *Vemurafenib* with the *Cobimetinib* treatment. OncoKB also reports an enhanced response to the combination of the *Dabrafenib* and *Trametinib* treatments. The concordance cannot be assessed for the other dimensions as there are no common variations in the data sources.

Since there are no variations that appear in more than one source, concordance for the **Oligoastrocytoma** cannot be calculated for any of the three dimensions (see Fig. 4).

Table 5 summarizes the results for the concordance in each disorder and clinical dimension.

4   Discussion

Our study has brought up the significant differences in terms of completeness and concordance between the information each data source provides. While the generic data sources contain both germline and somatic variations, the cancer-specific data sources only report somatic variations. However, there are two exceptions, namely, GWAS Catalog and CIViC. On the one hand, the GWAS Catalog is the only generic data source containing only germline variations. On the other hand, CIViC is the only cancer-specific source that includes both germline and somatic variations.

After studying the completeness at the schema level, we detected a significant heterogeneity in how the concept of clinical significance is represented. This heterogeneity leads to error-prone data integration processes and semantic interoperability issues. In this sense, academic works show how conceptual models and ontologies can mitigate this situation. These approaches help achieve a standardized representation of heterogeneous data, enhancing semantic interoperability [7,12].

The completeness at the data level results reflected significant differences in the variations each data source provides. Table 6 shows these results considering the variations for the three disorders. Cancer-specific data sources are the most complete for the dimensions under study: COSMIC for diagnostic, CIViC for prognostic, and OncoKB for predictive. Thus, they are the most valuable data sources for studying variations associated with cancer. However, generic data sources provide a small set of variations unavailable in cancer-specific data sources. This means that they must be considered for the study of cancer disorders in order to ensure that no relevant piece of information is missing. In this context, considering different data sources for applying precision oncology to offer patients accurate diagnoses and treatments is a must.

Table 6. Diagnostic, Prognostic, and Predictive completeness at the data level.

Dimension	ClinVar	Ensembl	COSMIC	CIViC	OncoKB
Diagnostic	1,13%	0,56%	98,86%	0,37%	0%
Prognostic	0%	0%	0%	100%	0%
Predictive	0%	0%	0%	2,75%	99,08%

Despite the few variations found in more than one data source (i.e., five variations), the analysis of the concordance dimension reported poor results, except in the diagnostic dimension of the Oligodendroglioma. These results complicate clinical decision-making processes because clinicians must work with information that is either contradictory or dispersed.

5   Conclusions

This paper studied the completeness and concordance of the information about cancer-related variations regarding three dimensions (i.e., the diagnostic, prognostic, and predictive dimensions). The reported results show that the information on cancer-associated variations is neither concordant nor complete. This makes achieving true precision oncology a nearly impossible task. Even worse, clinicians lack an easy and efficient way to determine whether there is complete/concordant information for variations identified in their patients. A limitation of this work is the amount of data available to study the concordance of the data. Thus, it is required to explore other types of cancer.

Future work will be oriented towards developing a web service that allows clinicians to obtain this information in an easy and efficient way. This will allow them to improve their decision-making processes.

References

1. Buniello, A., et al.: The NHGRI-EBI GWAS Catalog of published genome-wide association studies, targeted arrays and summary statistics 2019. Nucleic Acids Res. **47**(D1), D1005–D1012 (2018). https://doi.org/10.1093/nar/gky1120
2. Chakravarty, D., et al.: OncoKB: a precision oncology knowledge base. JCO Precis. Oncol. **1**, 1–16 (2017). https://doi.org/10.1200/PO.17.00011
3. Cunningham, F., et al.: Ensembl 2022. Nucleic Acids Res. **50**(D1), D988–D995 (2021). https://doi.org/10.1093/nar/gkab1049
4. Fokkema, I.F.A.C., et al.: LOVD v.2.0: the next generation in gene variant databases. Human Mutat. **32**(5), 557–563 (2011). https://doi.org/10.1002/humu.21438
5. Galperin, M.Y., Fernández-Suárez, X.M., Rigden, D.J.: The 24th annual nucleic acids research database issue: a look back and upcoming changes. Nucleic Acids Res. 45(D1), D1–D11 (2016). https://doi.org/10.1093/nar/gkw1188
6. Griffith, M., et al.: Civic is a community knowledgebase for expert crowdsourcing the clinical interpretation of variants in cancer. Nat. Genet. **49**, 170–174 (2017). https://doi.org/10.1038/ng.3774
7. Guizzardi, G.: Ontology, Ontologies and the "I" of FAIR. Data Intell. **2**(1–2), 181–191 (2020). https://doi.org/10.1162/dint_a_00040
8. Landrum, M., et al.: ClinVar: improving access to variant interpretations and supporting evidence. Nucleic Acids Res. **46**(D1), D1062–D1067 (2017). https://doi.org/10.1093/nar/gkx1153
9. Li, M., et al.: Standards and guidelines for the interpretation and reporting of sequence variants in cancer: a joint consensus recommendation of the association for molecular pathology, American society of clinical oncology, and college of American pathologists. J. Mol. Diagn. **19**, 4–23 (2017). https://doi.org/10.1016/j.jmoldx.2016.10.002
10. Li, X., Warner, J.L.: A review of precision oncology knowledgebases for determining the clinical actionability of genetic variants. Frontiers Cell Dev. Biol. **8**, 48 (2020). https://doi.org/10.3389/fcell.2020.00048
11. Pallarz, S., et al.: Comparative analysis of public knowledge bases for precision oncology. JCO Precis. Oncol. **3**, 1–8 (2019). https://doi.org/10.1200/PO.18.00371
12. Pastor, O., et al.: Using conceptual modeling to improve genome data management. Briefings Bioinform. **22**(1), 45–54 (2020). https://doi.org/10.1093/bib/bbaa100
13. Pipino, L., Lee, Y., Wang, R.: Data quality assessment. Commun. ACM **45**, 211–218 (2003). https://doi.org/10.1145/505248.506010
14. Rieke, D.T., et al.: Comparison of treatment recommendations by molecular tumor boards worldwide. JCO Precis. Oncol. **2**, 1–14. Wolters Kluwer (2018). https://doi.org/10.1200/PO.18.00098
15. Tate, J.G., et al.: COSMIC: the catalogue of somatic mutations in cancer. Nucleic Acids Res. **47**(D1), D941–D947 (2018). https://doi.org/10.1093/nar/gky1015
16. Wang, R.Y., Strong, D.M.: Beyond accuracy: what data quality means to data consumers. J. Manag. Inf. Syst. **12**, 5–33 (1996)
17. Weiskopf, N., Weng, C.: Methods and dimensions of electronic health record data quality assessment: enabling reuse for clinical research. J. Am. Med. Inform. Assoc. JAMIA **20**, 144–151 (2012). https://doi.org/10.1136/amiajnl-2011-000681

A Flexible Automated Pipeline Engine for Transcript-Level Quantification from RNA-seq

Pietro Cinaglia[1(✉)] and Mario Cannataro[2]

[1] Department of Health Sciences, Magna Graecia University of Catanzaro,
Catanzaro, Italy
cinaglia@unicz.it
[2] Data Analytics Research Center and Department of Medical and Surgical Sciences,
Magna Graecia University of Catanzaro, Catanzaro, Italy
cannataro@unicz.it

Abstract. The advances in massive parallel sequencing technologies (i.e., Next-Generation Sequencing) allowed RNA sequencing (RNA-seq). The analysis of RNA-seq data uses a large amount of computational resources, and it is very time-consuming. Usually, the processing is performed on a large set of samples, and it is convenient designing an automatic pipeline to eliminate the downtime. The pipelines represent an advantage, however these are difficult to customize, or to use outside the specific context for which they have been tested.

In this paper, we propose *FAPE* (Flexible Automated Pipeline Engine), a software platform to configure and to deploy automated pipelines. It models a pipeline based on a given template. The latter has a highly understandable and manipulable organization, to meet the operator's need for customization. In addition, a scientist may model an in-house custom pipeline able to execute all tools based on a command line interface (CLI). *FAPE* supports both parallel and iterative processes, in order to analyze whole datasets. We tested our solution on a pipeline for Transcript-level Quantification from RNA-seq, based on Hisat2, SamTools, and StringTie. It exhibited high robustness as well as inherent flexibility in supporting any pipeline modeled to specification. Furthermore, it has proven not to be expensive in terms of memory, and it does not introduce a significant latency during the execution, as compared to a pipeline executed through a shell-script program. In addition, the statement *parallel* of *FAPE* allowed during the test a reduction of the total elapsed time of $\sim 6.5\%$.
- https://github.com/pietrocinaglia/fape

Keywords: RNA-Seq · Pipeline · Transcriptome · Sequencing.

1 Introduction

The transcriptome is the whole set of RNA transcripts in a cell that allows to understand the mechanisms of development and disease [1]. It may be analyzed

© The Author(s), under exclusive license to Springer Nature Switzerland AG 2022
R. Guizzardi and B. Neumayr (Eds.): ER 2022 Workshops, LNCS 13650, pp. 45–54, 2022.
https://doi.org/10.1007/978-3-031-22036-4_5

to investigate the functional elements of a genome. Specifically, the transcription process [2] converts the DNA sequence into multiple transcript isoforms, for a specific gene. The isoforms are combinations of exons into a transcript (e.g., due to splicing) [3]. A cell switching from an isoform to another will have functional consequences; to give an example, an isoform is generally evaluated on cancer cells to study its outcomes, by comparing pathological and healthy patients [4]. In genomics, the raw data (e.g., biological sequence) is generally acquired by using Microarray and Next-Generation Sequencing (NGS) technologies. It is used for high-throughput large-scale RNA-level studies, and processed through bioinformatics algorithms [5]. However, the advances in massive parallel sequencing technologies (i.e., NGS) allowed RNA sequencing (RNA-seq). The latter may detect the RNA in a biological sample, also quantifying its presence [6].

A routine pipeline for RNA-seq [7,8] consists of the following main steps: (i) preprocessing of raw data (e.g., quality control) [9], (ii) mapping the reads to the reference genome or transcriptome (i.e., read alignment), (iii) identifying the transcripts expressed in a specimen of interest (i.e., expression quantification), (iv) expression quantification, and (v) differential expression analysis. The expression quantification concerns the gene-level quantification and isoform-level quantification, while the differential expression analysis concerns the gene-level and isoform-level evaluation.

Potentially, an RNA-seq scenario has several optimal methods, so long as based on specifically best-practices [10]. For instance, the transcript-level quantification may be performed by using Hierarchical Indexing for Spliced Alignment of Transcripts 2 (Hisat2) [11] and Stringtie [12]. Hisat2 performs the short read mapping by implementing an aligner based on Bowtie [13]. Stringtie is an assembler for RNA-Seq alignments able to assemble and to quantify the expression levels by performing the transcriptome reconstruction and the abundance estimation, simultaneously. It identifies the path of the heaviest coverage on a network built ad-hoc, and it evaluates the maximum flow to estimate abundance for this one [14]. Similarly, the same pipeline, may be performed by using Cufflinks [15] for isoform-level quantification, in place of Stringtie. It also quantifies the related feature expressions for each assembled transcript.

Pertea et al. [16] presented a pipeline for transcript-level expression analysis of RNA-seq experiments based on HISAT2, and StringTie. It described the steps useful for read mapping, transcripts identification, transcript-level quantification, as well as to process the differentially expressed genes and transcripts. Similarly, Trapnell et al. [17] proposed a pipeline for differential gene and transcript-level expression analysis of RNA-seq experiments, based on TopHat [18] and Cufflinks. Both the pipelines need to be executed though a shell, manually, or to be implemented in a shell-script program, alternatively. For instance, Spinozzi et al. [19] presented an automatic RNA-Seq pipeline based on a main script developed in Python language. It calls a series of activities (e.g., alignment), in turn developed as script. Therefore, programming skills are requested to develop a novel activity, as well as the main script must be updated for each new activity, in order to support it. Similarly, Srivastava et al. [20] proposed an automatic pipeline for network analysis from RNA-Seq, based on R language. Data processing related to RNA-seq consists of a large use of resources exploited, e.g., for mapping of millions of reads, for expression analysis of thousand genes and transcripts, for statistical analysis. For this reason, RNA-seq data analysis is usually performed by using platforms for large scale cluster/cloud computing. In this context, the data processing based on an automatic pipeline represents an advantage. However, the latter may be difficult to customize, or to use outside the specific context for which it has been tested [21].

In this paper, we propose *FAPE* (Flexible Automated Pipeline Engine), a software platform to configure and deploy automated pipelines for Transcript-level Quantification from RNA-seq. It allows the execution of a given pipeline modeled on a specific template. The latter has a highly understandable and manipulable organization, to meet the operator's need for customization. In addition, a scientist may model an in-house custom pipeline able to execute all tools based on a command line interface (CLI). We focused the attention on Transcript-level Quantification from RNA-seq, even other applications could concern other analysis, or data.

The rest of the paper is organized as follows. Section 2 presents the design, the architecture, and the implementation of *FAPE*. Section 3 describes both the testing pipeline for Transcript-level Quantification from RNA-seq and the related analysis. Finally, Sect. 4 concludes the paper, presenting also the future works.

2   Methods

Our solution (i.e., *FAPE*) is an in-house engine based on pipelines modeled in according to a specific template. This section presents the methodologies related to the engine, by focusing on an in-house template designed for the Transcript-level Quantification from RNA-seq.

We developed *FAPE* in Python language (version 3). It supports both parallel and iterative processes, in order to analyze whole datasets. It allows analyzing more than one data, without being restarted or reconfigured. Furthermore, its steps can be related to each other to exchange data and resources. Each step executes a command, and a set of commands may be grouped within a block. More blocks can be linked between them, or used independently, to model the pipeline.

FAPE consists of the following main components:

- *template renderer* is able to parse the template, and to configure it based on input parameters;
- *engine* orchestrates each pipeline's block into an end-to-end process, according to the defined steps;
- *monitor* provides the information about the processes (e.g., status, identifier, standard output, standard error);

- *modules* inspects and configures the dependencies (i.e., third-party software tools).

Foremost, *FAPE* needs a template in input representing the pipeline. It loads this one, by also inspecting the parameters defined inside. A parameter may have a default value that is used if not specified by the user. For instance, the default value for the number of threads is equal to the number of cores (or virtual cores) on the host's central processing unit (CPU); for more CPUs is the sum of all cores. However, *FAPE* uses the CLI to ask for the missing parameters. Subsequently, the template is inspected to retrieve the list of modules called by the commands, by also checking if these are available on the current host. Each module is configured by *FAPE* through the retrieved parameters, then the execution of pipeline is started. The *engine* orchestrates each pipeline's block into an end-to-end process, according to the steps defined for it. When a block is loaded for the execution, its statement is evaluated. *FAPE* supports three types of statements:

- *foreach* provides an easy way to execute the set of commands for each *object*, in sequence;
- *parallel* allows executing the set of commands in parallel;
- *none* (or empty statement) executes the steps sequentially, without taking into account the presence of multiple samples.

Note that *foreach* and *parallel* can also be used together. In this case, a balancer handles the threads, by dividing equally these among the *objects*. If the number of threads is less than the objects, then *FAPE* applies a First In First Out (FIFO) queue, in accordance to the order of retrieving.

In addition, the *engine* handles both the standard output (stdout) and the standard error (stderr) messages of each child process in execution. The stdout is a descriptor where each process writes its output. Similarly, the stderr is a descriptor where each process writes its error messages. *FAPE* supports the stdout and the stderr according to the Portable Operating System Interface for uniX (POSIX) standard.

Figure 1 reports the described components, and it summarizes the basis of the functioning of *FAPE*.

As mentioned, *FAPE* needs a template in input representing the pipeline. The latter must be designed by using the JavaScript Object Notation (JSON) syntax-compatible file format, in accordance to *FAPE* specification.

A pipeline consists of the following main parts:

- the header provides information about the pipeline (e.g., name, description, and author).
- the blocks represent the set of steps on which the macro-process have to be built.
- the steps are the subunits of a block. We may consider a step as a child-process within a block.

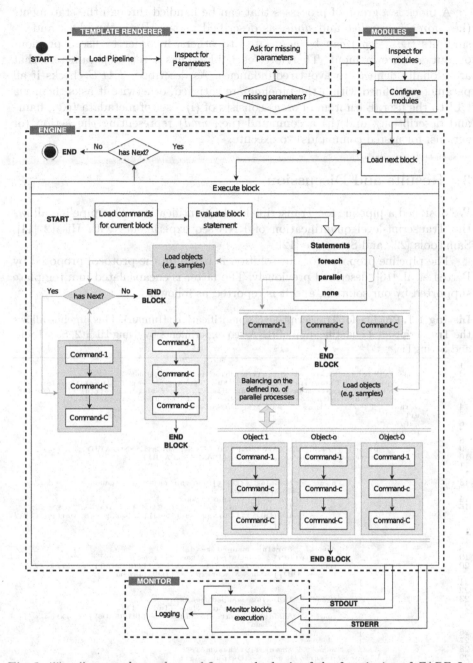

Fig. 1. The diagram shows the workflow at the basis of the functioning of *FAPE*. It reports the described components: *template renderer*, *engine*, *monitor*, and *modules*, as well as the statements (i.e., *foreach*, *parallel*, and *none*), and the procedures related to these.

A block is a group of processes that can be handled through the statements (i.e., *foreach*, *parallel*, and *none*, as described above). Both the block and its subunits (i.e., *steps*) can be associated to others, in order to use a previous output as an own input. Therefore, each block has a set of parameters that are globally defined, to avoid redundancies. As described, *FAPE* checks if all parameters defined within the pipeline are satisfied, otherwise it asks them via CLI to the user. Each item of *steps* consists of (i) a set of metadata (e.g., name and description), and (ii) a command (i.e., *cmd*) representing the action (or actions, for nested commands) to execute.

3   Results and Discussion

We designed a pipeline for Transcript-level Quantification. This pipeline allows the Transcript-level quantification of RNA-seq experiments with Hisat2 [11], SamTools [22], and StringTie [12].

The pipeline proposed for our solution is based on the protocol proposed by Pertea et al. [16], described previously. The latter is encapsulated in a template supported by our solution, and it is reported as follows.

Listing 1.1. Pipeline for Transcript-level Quantification (Human). This pipeline allows the Transcript-level quantification of RNA-seq experiments by using Hisat2, SamTools, and StringTie.

```
1  {
2      "name": "Transcript-level Quantification pipeline (Human)",
3      "description": "This pipeline allows the Transcript-level Quantification of
              ↪ RNA-seq experiments by using Hisat2, SamTools, and StringTie. Output
              ↪ enriched with coverage data for the reference transcripts.",
4      "author": "Pietro Cinaglia and Mario Cannataro",
5      "url": "",
6      "help": "",
7      "pipeline": {
8          "1": {
9              "params": "THREADS,SAMPLES_PATH,GENOME,GRCh38_ANNOTATED",
10             "statement": "foreach OBJ in {{SAMPLES_PATH}}",
11             "steps": [
12                 {
13                     "name": "Mapping {{OBJ}} on the reference genome (i.e.,
                            ↪ GENOME) [.sam]",
14                     "description": "- Output: .sam",
15                     "cmd": "./modules/hisat2 -p {{THREADS}} --dta -x {{GENOME}}
                            ↪ -1 {{OBJ}}_1.fq.gz -2 {{OBJ}}_2.fq.gz -S workspace/{{
                            ↪ OBJ}}.sam"
16                 },
17                 {
18                     "name": "Sorting mapped reads",
19                     "description": "- Output: .bam (sorted)",
20                     "cmd": "./modules/samtools sort -@ {{THREADS}} workspace/{{
                            ↪ OBJ}}.sam -o workspace/{{OBJ}}.bam"
21                 },
22                 {
23                     "name": "Indexing mapped reads",
24                     "description": "- Output: .bam (indexed)",
25                     "cmd": "./modules/samtools index -@ {{THREADS}} workspace/{{
                            ↪ OBJ}}.bam"
26                 },
27                 {
28                     "name": "Transcript assembly",
29                     "description": "- Output: .gtf",
30                     "cmd": "./modules/stringtie -p {{THREADS}} workspace/{{OBJ}}
                            ↪ .bam -l {{OBJ}} -G {{GRCh38_ANNOTATION}} -o workspace
                            ↪ /{{OBJ}}.gtf --rf -A workspace/{{OBJ}}_gene_abundance
                            ↪ .tab"
31                 }
32             ]
33         },
34         "2": {
35             "params": "SAMPLES_LIST",
```

```
36    "steps": [
37        {
38            "name": "Merging all transcripts",
39            "description": "- Output: merged.gtf",
40            "cmd": "./modules/stringtie --merge -p {{THREADS}} -G {{
              ↪ GRCh38_ANNOTATION}} -o workspace/merged.gtf {{
              ↪ SAMPLES_LIST}}"
41        }
42    ]
43   },
44   "3": {
45       "params": "",
46       "statement": "foreach OBJ in {{SAMPLES_PATH}}",
47       "steps": [
48           {
49               "name": "Estimate transcript-level abundances (ta) in {{OBJ}
                 ↪ }",
50               "description": "- Output: .ta.gtf, and Ballgown folder (-B)
                 ↪ .",
51               "cmd": "./modules/stringtie -p {{THREADS}} -e -B -G
                 ↪ workspace/merged.gtf workspace/{{OBJ}}.bam -o
                 ↪ workspace/{{OBJ}}.tla.gtf"
52           }
53       ]
54    }
55   }
56 }
```

For testing, we used an existing dataset of RNA-Seq reads (ftp://ftp.
sra.ebi.ac.uk/vol1/fastq/ERR188/ accessed on 03 July 2022). The samples' directory contains the following paired-end RNA-seq reads as compressed
FastQ (.fq.gz): ERR188245, ERR188428, ERR188337, ERR188401, ERR188257,
ERR188383. A paired-end FastQ consists of two files for each sample; to give an
example, ERR188245 is composed by $ERR188245\_1.fq.gz$ and $ERR188245\_2.$
$fq.gz$, respectively. In our pipeline, the $SAMPLES\_PATH$ represents the path
for the samples' directory. For testing, we used the HISAT2 whole genome index
from the Ensembl Genome Reference Consortium Human Build 38 (Human
GRCh38, ftp.ccb.jhu.edu/pub/infphilo/hisat2/data/hg38_tran.tar.gz
accessed on 03 July 2022). $FAPE$ has been executed on a 64-bit computer running Linux (Intel Xeon 3.40 GHz, 8 processing cores, and 16 GB of RAM).

The RNA-seq analysis begun by mapping the data on the reference genome
by using Hisat2, in order to identify the related genomic positions. The output,
provided by Hisat2, was sorted and indexed by chromosome through SamTools.
Subsequently, StringTie processed the following protocol: (i) transcript assembly
for each output, (ii) merging all outputs, (iii) estimating Transcript-level abun-
dances on each output, against all merged outputs. It refers to the step 1.4 ($step4$
of $block1$), step 2 ($block2$, having only one step), and step 3 ($block3$, having only
one step), within our pipeline respectively.

Our solution exhibited high robustness, as well as inherent flexibility in supporting any pipeline modeled to specification. Furthermore, it has proven not to
be expensive in terms of memory: 14.62 MB of memory for the basic usage ($mon-
itor$ activated, and no running processes from third-party software tools), +0.24
MB of memory (on basic usage) to build the environment for pipeline loading
(i.e., $template renderer$) and modules checking (i.e., $modules$), while the $engine$
orchestrated the processes by using only +0.58 MB of memory (on basic usage).
Other resources are related to the third-party software tools (i.e., Hisat2, Samtools, StringTie). Furthermore, we evaluated the elapsed time for the described

pipeline executed by using both our solution and a shell-script program, in order to study any latencies.

Table 1 reports the elapsed times related to the pipeline execution *FAPE* with the statement *parallel*, *FAPE* in sequence mode (statement set to *none*), and shell-script program. Results showed that *FAPE* does not introduce a significant latency compared to a shell-script program. Furthermore, we tested the same pipeline by adding the statement *parallel*, on *FAPE*. In this test, results showed a reduction of total elapsed time of $\sim$ 6.5%. We assumed that the motivation may be related to the real use of the CPU in the multithreading of the individual modules. To give an example, Samtools, set to 8 cores, does not really use the 8 cores at 100% during the whole processing. The statement *parallel* overloads the CPU usage with multiple processes concurrently, therefore these take advantage of times when a single execution does not fully utilize the CPU.

Table 1. Elapsed times related to the pipeline executed by *FAPE* with the statement *parallel*, *FAPE* in sequence mode (statement set to *none*), and shell-script program. Times are reported in minutes.

	FAPE (parallel)	FAPE	Shell-Script program
Mapping, Sorting, and Indexing	*not estimable for single sample*	$\sim$70/sample	$\sim$70/sample
Transcript assembly	*not estimable for single sample*	$\sim$15/sample	$\sim$15/sample
Merging all transcripts	$\sim$20	$\sim$20	$\sim$20
Estimating transcript-level abundances	*not estimable for single sample*	$\sim$15/sample	$\sim$15/sample
Elapsed time for all 6 samples	$\sim$692	$\sim$740	$\sim$740

4   Conclusions and Future Work

This paper presented a Flexible Automated Pipeline Engine (i.e., *FAPE*), tested for Transcript-level Quantification from RNA-seq, based on Hisat2, SamTools, and StringTie. Our solution turned out to be flexible, in supporting any pipeline modeled in accordance to the defined specifications for templates. It exhibited high robustness as well as inherent flexibility. Furthermore, it has proven not to be expensive in terms of memory, without significant latencies during execution as compared to a pipeline executed through a pure shell script. In our test, the statement *parallel* of *FAPE* allowed a reduction of the total elapsed time of $\sim$6.5%.

Future works could concern the modeling of other genomics pipelines, parallelism improvements, and the implementation of a Graphical User Interface (GUI), as well as the porting of existing bioinformatics solutions [23,24] as *FAPE*'s template.

References

1. Yang, I.S., Kim, S.: Analysis of whole transcriptome sequencing data: workflow and software. Genomics Inform. **13**(4), 119–125 (2015)

2. Li, J., Liu, C.: Coding or noncoding, the converging concepts of RNAs. Front. Genet. **10**, 496 (2019)
3. Thomas, Q.A., et al.: Transcript isoform sequencing reveals widespread promoter-proximal transcriptional termination in Arabidopsis. Nat. Commun. **11**(1), 2589 (2020)
4. Nielsen, M., et al.: Transcription-driven chromatin repression of Intragenic transcription start sites. PLoS Genet. **15**(2), e1007969 (2019)
5. Cinaglia, P., Guzzi, P.H., Veltri, P.: Integro: an algorithm for data-integration and disease-gene association. In: 2018 IEEE International Conference on Bioinformatics and Biomedicine (BIBM), pp. 2076–2081 (2018)
6. Denoeud, F., et al.: Annotating genomes with massive-scale RNA sequencing. Genome Biol. **9**(12), R175 (2008)
7. Creason, A., et al.: A community challenge to evaluate RNA-seq, fusion detection, and isoform quantification methods for cancer discovery. Cell Syst. **12**(8), 827–838 (2021)
8. Haas, B.J., et al.: De novo transcript sequence reconstruction from RNA-seq using the Trinity platform for reference generation and analysis. Nat. Protoc. **8**(8), 1494–1512 (2013)
9. Yang, X., et al.: HTQC: a fast quality control toolkit for Illumina sequencing data. BMC Bioinform. **14**, 33 (2013)
10. Conesa, A., et al.: A survey of best practices for RNA-seq data analysis. Genome Biol. **17**, 13 (2016)
11. Kim, D., Paggi, J.M., Park, C., Bennett, C., Salzberg, S.L.: Graph-based genome alignment and genotyping with HISAT2 and HISAT-genotype. Nat. Biotechnol. **37**(8), 907–915 (2019)
12. Pertea, M., Pertea, G.M., Antonescu, C.M., Chang, T.C., Mendell, J.T., Salzberg, S.L.: StringTie enables improved reconstruction of a transcriptome from RNA-seq reads. Nat. Biotechnol. **33**(3), 290–295 (2015)
13. Langmead, B., Trapnell, C., Pop, M., Salzberg, S.L.: Ultrafast and memory-efficient alignment of short DNA sequences to the human genome. Genome Biol. **10**(3), R25 (2009)
14. Kim, D., Pertea, G., Trapnell, C., Pimentel, H., Kelley, R., Salzberg, S.L.: TopHat2: accurate alignment of transcriptomes in the presence of insertions, deletions and gene fusions. Genome Biol. **14**(4), R36 (2013)
15. Trapnell, C., et al.: Transcript assembly and quantification by RNA-seq reveals unannotated transcripts and isoform switching during cell differentiation. Nat. Biotechnol. **28**(5), 511–515 (2010)
16. Pertea, M., Kim, D., Pertea, G.M., Leek, J.T., Salzberg, S.L.: Transcript-level expression analysis of RNA-seq experiments with HISAT, StringTie and Ballgown. Nat. Protoc. **11**(9), 1650–1667 (2016)
17. Trapnell, C., et al.: Differential gene and transcript expression analysis of RNA-seq experiments with TopHat and cufflinks. Nat. Protoc. **7**(3), 562–578 (2012)
18. Trapnell, C., Pachter, L., Salzberg, S.L.: TopHat: discovering splice junctions with RNA-seq. Bioinformatics **25**(9), 1105–1111 (2009)
19. Spinozzi, G., Tini, V., Adorni, A., Falini, B., Martelli, M.P.: ARPIR: automatic RNA-seq pipelines with interactive report. BMC Bioinform. **21**(Suppl 19), 574 (2020)
20. Srivastava, H., Ferrell, D., Popescu, G.V.: NetSeekR: a network analysis pipeline for RNA-seq time series data. BMC Bioinform. **23**(1), 54 (2022)

21. Wratten, L., Wilm, A., Göke, J.: Reproducible, scalable, and shareable analysis pipelines with bioinformatics workflow managers. Nat. Methods **18**(10), 1161–1168 (2021)
22. Danecek, P., et al.: Twelve years of SAMtools and BCFtools. GigaScience **10**(2), giab008 (2021)
23. Cinaglia, P., Cannataro, M.: Forecasting COVID-19 epidemic trends by combining a neural network with rt estimation. Entropy (Basel) **24**(7), 929 (2022)
24. Cinaglia, P., Tradigo, G., Cascini, G.L., Zumpano, E., Veltri, P.: A framework for the decomposition and features extraction from lung dicom images. In: Proceedings of the 22nd International Database Engineering & Applications Symposium, pp. 31–36. IDEAS 2018, Association for Computing Machinery (2018)

An Initial Empirical Assessment of an Ontological Model of the Human Genome

Alberto García S.[1]([✉]), Anna Bernasconi[1,2], Giancarlo Guizzardi[3,4],
Oscar Pastor[1], Veda C. Storey[5], and Mireia Costa[1]

[1] Universitat Politècnica de València, Valencia, Spain
{algarsi3,pastor,micossan}@pros.upv.es, abernas@upvnet.upv.es
[2] Politecnico di Milano, Milan, Italy
anna.bernasconi@polimi.it
[3] Free University of Bozen-Bolzano, Bolzano, Italy
Giancarlo.Guizzardi@unibz.it
[4] University of Twente, Twente, The Netherlands
g.guizzardi@utwente.nl
[5] Georgia State University, Atlanta, GA, USA
vstorey@gsu.edu

Abstract. Conceptual modeling is used to model application domains for which an information system is needed. One of the most complex domains to which conceptual modeling has been applied is that of the human genome. Due to its complexity, its understanding is often left to domain experts. Conceptual models represent genomics-related concepts, with various purposes, including domain clarification or data structures design for facilitating data integration. However, traditional conceptual models, which might be expressed, for example, with UML, may not be appropriate for properly explaining such a complex domain, thus requiring an additional layer to ground the model on well-accepted ontological foundations. To achieve this result, an "ontological unpacking" method has been proposed that uses OntoUML as a visual formalism. In this research, we carry out an empirical study to compare the two mentioned representations. The study involved a small group of participants, who responded to a set of questions by reading either a UML model or its related OntoUML unpacked version; the results enabled us to assess their understanding of the domain. We aim to initiate a practical evaluation framework to assess the effectiveness, efficiency and user beliefs of models derived by ontologically unpacking traditional conceptual models. The results of the analysis provide the basis for a broader assessment.

Keywords: Empirical evaluation · Ontological unpacking ·
Conceptual model · Human genome

A. García S. and A. Bernasconi should be regarded as joint first authors.

R. Guizzardi and B. Neumayr (Eds.): ER 2022 Workshops, LNCS 13650, pp. 55–65, 2022.
https://doi.org/10.1007/978-3-031-22036-4_6

1 Introduction

Genomic science is a complex interdisciplinary domain, whose understanding is so far accessible only to researchers with a strong background in biology and genetics. Its interpretation becomes problematic also because there has been a lack of effort to translate its mechanisms into modeling languages that are more easily understandable by computer scientists. Computer science traditionally employs modeling languages such as UML, i.e., a standard graphical modeling language that allows designers to create conceptual models. Instead, OntoUML is an ontologically well-founded language for Ontology-driven Conceptual Modeling, built as a UML extension based on the Unified Foundational Ontology [8]. OntoUML supports modelers in systematically making ontologically consistent representation choices, and thus, making explicit the ontological nature of the elements represented.

We previously created a method of ontological analysis that reveals the ontological foundation of the information represented in a conceptual model. The method, called 'ontological unpacking', allows the modeler to unfold and explain previously existing UML models, transforming them into a corresponding OntoUML version. A previous effort has been successfully performed on the viral sequences domain [9] in order to improve semantic interoperability.

We postulate that ontologically unpacked models provide a clear and understandable representation of a complex domain, such as genomics, even though it might require considerable effort to learn OntoUML, which is necessary to perform the unpacking. To investigate, we here conduct an initial experiment where students without previous biological knowledge are given competency questions regarding a conceptual model, using either a UML model or its corresponding OntoUML model, obtained as a result of an ontological unpacking procedure. The experiment is carried out using a portion of a conceptual schema of the human genome as a complex domain [5,6]. Specifically, we consider the part describing human metabolic pathways. The original schema was conceived using UML; in our recent work [4] we performed an ontological analysis exercise, producing the corresponding OntoUML version. These two models are object of the experiment thereon described. We formulate a set of research questions aimed at understanding if OntoUML delivers better quality models than UML. Our research questions can be translated into formal metrics according to ISO 25000 (i.e., effectiveness, efficiency and user beliefs).

Ontology driven conceptual modeling has previously been compared to traditional conceptual modeling in [16]. Here, we do not compare different languages or paradigms of modeling; instead, we compare the capability of different models to completely and unambiguously represent a domain, serving the intended purpose of explaining that domain to a non-expert user working in it for the first time. Empirical studies of conceptual modeling applications have been performed on tools [7] also related to genomics [2], measuring the understandability of modeling artifacts [11,12]. This paper describes our initial experiment and discusses a number of lessons learned. Future work will include further experimentation and statistical analysis to assess the use of ontological unpacking.

2   Background

The first Conceptual Schema of the Human Genome was proposed in 2011 [14] by the Research Center on Software Production Methods (PROS) at the Polytechnic University of Valencia. Since then, several extensions have been produced [5,15]. The current version of the schema is a map of concepts and relationships grouped into different genomic knowledge modules, called the Conceptual Schema of the Genome v3 (CSG) [6]. Its UML schema includes five modules, describing the structure of the human genome, protein synthesis, changes in the sequence referring to a reference sequence, information and sources related to the elements of the conceptual schema, and human metabolic pathways. Prior work on ontological unpacking [4] of the schema focused on a relevant portion of the last view; that is, on metabolic pathways. We sought to understand the impact of ensuring ontological clarity in the concepts employed.

The original UML pathway schema presents 19 entity classes, with six generalizations, two aggregations, one self-relation, and three normal relations. We also have one integrity constraint. The unpacked OntoUML schema has 26 entities, of which 17 are from UFO-A [10] (including kinds and subkinds, collectives and categories, phase/roleMixins) and 9 from UFO-B [1], including 5 events and 4 historicaRolelMixins.

Different relationships include 11 generalizations, three aggregations, one composition, and four other regular ones, covering several relationship stereotypes; namely, «creation», «termination», «memberOf», «participational», and «historicalDependence». Here we do not explain OntoUML stereotypes but we refer the interested readers to [8].

3   Methodology

We carried out an empirical assessment by designing a study to answer three fundamental research questions:

RQ1: *Do subjects benefit from a better understanding of a complex domain with OntoUML rather than with UML?*
RQ2: *Do subjects answer competency questions faster with OntoUML than with UML?*
RQ3: *Do subjects have more positive beliefs after using OntoUML rather than UML?*

The experimental design is based on the one described in [16], thus divided into four steps, namely: variable development, subject selection, experimental design type, and instrumentation.

Variable Development. Based upon our research questions, our independent variable is the modeling language, with two possible treatments, UML and OntoUML. The dependent variable is the quality of the models, observed using three dimensions, captured with different metrics:

- *Effectiveness*: measured through the percentage of correct true/false answers given to competency questions.
- *Efficiency*: measured through the time to answer to competency questions.
- *User beliefs* – divided into the three sub-dimensions *perceived usefulness* (PU), *perceived ease of use* (PEOU), and intention to use (ITU): measured through 1-to-5 Likert scale questionnaires.

Subject Selection. The experiment was performed at the Polytechnic University of Valencia in two classes of the fourth year of the Computer Science curriculum. There were 20 participants aged 22 to 30. Given the small sample size, this can be considered as a quasi-experiment. Only the previous IT background of the participants was taken into account, since none of the subjects declared any previous experience with biological/genomic topics.

Experimental Design Type. Participants all together filled a demography survey, received an introduction to the two involved modeling languages and were tested on them. Then they were divided into four groups. Figure 1 shows the experimental setup given to each of them. We created two questionnaires, Q1 and Q2, that are equivalent in terms of difficulty and coverage of modeling constructs/topics. Participants were asked to answer Q1 and Q2 in different orders, using alternatively: 1) the original UML-based CSG pathway view, or 2) the ontologically unpacked OntoUML-based CSG pathway view. In this way, all the four possibilities (questionnaire Q1 or Q2 answered with UML or OntoUML schemata) were covered, overcoming possible biases due to the order of issuing. We performed random assignment of Q1 and Q2 to the subjects.

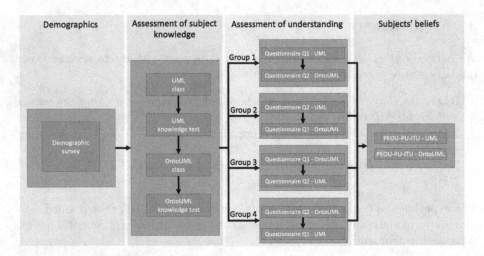

Fig. 1. Experimental design overview. [PEOU: Perceived ease of use; PU: Perceived usefulness; ITU: Intention to Use]

Table 1. Questionnaire competency questions

Quest.	Group*	ID	Competency questions
Q1	Entities	1	Polymers are composed of other polymers
		2	The internal structure of any polymers is homogeneous
		3	The internal structure of basic biological entities and polymers is the same
	Events	4	Processes are limited in time
		5	Pathways must be composed of other pathways
		6	A process can be decomposed into other events
	Interact.	7	Every biological entity must participate in at least one process
		8	Biological entities can take part in pathways
		9	A protein can take the roles of input, output, and regulator in the same process
Q2	Entities	10	Some polymers are composed of nucleotides
		11	Every enzyme is a polymer
		12	Some basic biological entities can be polymers also
	Events	13	Every event must have a preceding event
		14	Pathways can be composed of other pathways
		15	Events occur in a specific time interval
	Interact.	16	Biological entities can be created and destroyed as a result of a process
		17	Biological entities can participate in multiple processes
		18	A protein can take the role of input in different processes

Instrumentation. The experiment was conducted using the PoliFormat platform (https://poliformat.upv.es/), on which the students had a personal login access. All the materials used for introducing the topics to the classes and to assess their understanding are provided as supplementary material on a Zenodo repository [3]. In the remainder of the text, we refer to specific handouts using the same acronyms as in the repository.

Demographics. As shown in the first block of Fig. 1, the demographic survey consisted of eight questions (see file DS) aimed to grasp a more complete picture of the participants group to better interpret the results.

Assessment of Subject Knowledge. As shown in the second block of Fig. 1, training on both UML and OntoUML was offered to all study participants (see files SK1 and SK2), aiming to eliminate all possible differences due to the background knowledge of the participants. Each training session lasted 45 min and was supported by a slides presentation, including theory and practical examples taken from domains not related to the one in the experiment. After each training session, participants answered to a questionnaire testing their understanding of example models (see files SK3 and SK4).

Assessment of Understanding. As shown in the third block of Fig. 1, the participants were divided into four groups, which answered the same sets of questions by using two different models: 1) the original UML model (see file UA5); 2) ontologically unpacked OntoUML model (see file UA6). The questions were provided by Biology expert collaborators, as they deemed them relevant for the domain. In this way, questions were guaranteed to be independent with respect to the models. After processing the statements provided by experts, we composed three groups of questions, respectively targeting Entities, Events, or Interactions between entities and events. This pre-processing allowed the composition of questionnaires Q1 and Q2 in a balanced way with respect to the models' interpretation challenges. Table 1 shows the sets of questions divided by questionnaire number and by group. The different versions of the questionnaires for the 4 groups are available in handouts UA1–UA4. A pilot run was performed before the experiment with expert collaborators to ensure that the task had the appropriate level of complexity.

Subjects' Beliefs. As shown in the fourth block of Fig. 1, all participants finally answered two surveys of 16 questions to assess their beliefs, in terms of PU (8 questions), PEOU (6 questions), and ITU (2 questions) according to Method Adoption Model (MAM, [13]), measured using a Likert scale.

4   Results

Results obtained by running the experiment with the selected participants follow.

Demographics. The involved students declared a Grade Point Average of about 8/10. Their working experience was very heterogeneous. Most participants had <1 year work experience, whereas two had worked longer. Their demographic survey results revealed that the participants: i) knew and had previously used UML; ii) had no previous experience with OntoUML; and iii) were not knowledgeable in the observed domain.

Assessment of Subject's Knowledge. Eighteen students successfully passed the two UML and OntoUML tests with >75% correct answers in both tests. One participant did not obtain the sufficient threshold in OntoUML (62.5%); another did not pass either tests (50% and 57.14%). These two students received an additional class and were dedicated additional time for answering any questions they had. We ensured that sufficient understanding was reached before proceeding to the next stages.

Effectiveness of Treatment. Figure 2A shows, for each question, the percentage of correct answers received by participants using either UML or OntoUML. Questions are grouped by category. It can be observed that: i) Entities-related questions were answered correctly by 68.33% of participants using UML and by 76.67% using OntoUML; ii) Events-related questions were answered correctly respectively by 56.67% (UML) and 83.33% (OntoUML); and iii) Interactions-related questions were answered correctly respectively by 58.33% (UML) and 56.67% (OntoUML).

Efficiency of Treatment. Figure 2B shows, for each question, the working mean times (measured in seconds) spent by the participants to provide answers, using either UML or OntoUML. Questions are grouped by category. Questions answered with the OntoUML model took longer than questions answered with

UML. Specifically: i) questions related to Entities and Interactions required subjects approximately 30 s longer to answer; ii) the difference decreases to approximately 20 s for Event-related questions; iii) times required to answer UML-based questions showed a higher variability than those for OntoUML-based questions. For example, the time required to answer Entities-related questions in UML ranged from 63 to 89 s (26 s difference) and from 95 to 109 s (14 s difference) in OntoUML.

Subjects' Beliefs. Figure 3A shows, for each question of the MAM (grouped by sub-dimension of the user belief), the partition of respondents who strongly disagreed, disagreed, was neutral, agreed, or strongly agreed with the provided statement. The same structure is used in Fig. 3B for OntoUML. From the results, it could be observed that: i) subjects perceived that UML is much easier to use (the average in PEOU questions with UML scored 0.83 more than with OntoUML); ii) subjects perceived that UML is more useful (difference of PU averages 0.38); iii) if subjects had to choose which language to use in genomics, they would prefer UML by a substantial margin (difference of ITU averages 0.9).

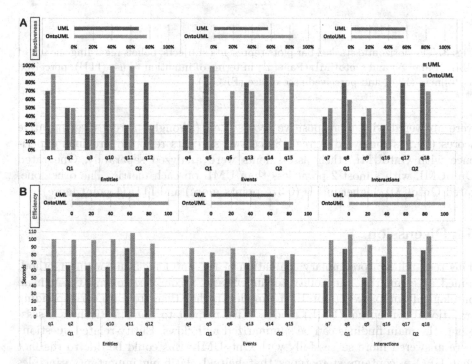

Fig. 2. Panel A: Barplot of correct answers % given by participants grouped by question number and organized by group. Panel B: Barplot of seconds employed to answer questions, grouped by question number and organized by group.

OntoUML results regarding user beliefs generally scored worse than UML results. The single OntoUML questions regarding perceived usefulness (PU)

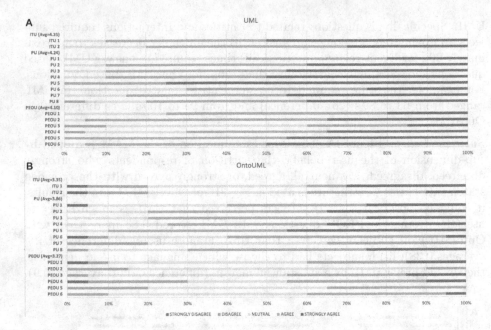

Fig. 3. The horizontal stacked bar plots represents subjects' beliefs regarding the use of UML (Panel A) and OntoUML (Panel B) in terms of intention to use (ITU), perceived usefullness (PU), and perceived ease of use (PEOU).

were answered with more positive scores (i.e., (strongly) agree) than negative scores (i.e., (strongly) disagree). Since two subjects reported previous experience with OntoUML, their assessments were analyzed separately. One rated OntoUML with almost 2 point less than UML on each metric. The other one, rated OntoUML higher in PU (0.375 points more) and ITU (1 point more).

5   Discussion

This research is a preliminary evaluation of the ontological unpacking method, aimed at comparing the ability to understand a complex domain through an ontologically unpacked (OntoUML) model, rather than from its corresponding traditional conceptual (UML) model. With respect to the Effectiveness assessment, the main findings can be summarized as follows. Entity-related questions were answered more successfully with OntoUML; this could be due to the fact that UFO-A contains stereotypes that helped clarifying important principles (such as rigidity). Events-related questions were also answered more successfully with OntoUML, showing an even more apparent difference; this suggests that the ontological foundation of events presented in the UFO-B fragment may have helped participants to capture relevant details regarding event-related information. Questions related to the Interaction between events and entities were instead answered more successfully with UML (2% difference). Further comments can be formulated by analyzing the results of specific questions:

- *Temporality of events.* One of the main purposes of Conceptual Modeling is making implicit concepts explicit. From a biological perspective, it is clear that events are limited in time. However, in the UML version of the considered model, the temporal limitations of a process are implicit. From the ontological unpacking method, which also commits to UFO-B, such information was extracted and explicitly represented by means of the «event» stereotype. Results of questions Q4 (OntoUML: 90% of participants answered correctly, UML: 30%) and Q15 (OntoUML: 80%, UML: 10%), which grasped this aspect, were thus expected.
- *Mereology of events.* The UML version of the considered model provides a simple explanation of the participation of entities in the processes. OntoUML, instead, provides a more complex and detailed explanation. Note that processes in genomics are considered chemical compounds that can be divided further. Question Q6 highlighted that the UML model left the individual participation of chemical compounds in reactions implicit. As a result, the question was answered with a higher score using OntoUML (70%) instead of UML (40%). However, representing how chemical reactions are decomposed into smaller events (which capture the individual participation of each chemical compound) increased the overall complexity. This may have confused respondents of Q17 (which concerns participation in multiple processes), who ended up scoring 20% with OntoUML and 80% with UML.
- *The rigidity principle.* A significant difference is observed in Q16 (OntoUML 90% vs. UML 30%), possibly due to the capability of OntoUML to express the «phase» stereotype, exploiting the principle of rigidity [8]. This principle makes explicit the fact that chemical compounds and biological-related substances are created and destroyed as a result of chemical reactions.

Thus for *RQ1*, we can conclude that OntoUML was more effective in conveying the genomics domain to the study participants, even if for some elements, the simplicity of the UML representation still achieved the desired outcome.

Regarding the Efficiency assessment, the initial expectation suggested that a complex domain explained through a more complete and explicit model would translate into shorter answering times. However, this preliminary study suggested that OntoUML required instead more time to participants, in order for them to be able to answer questions based on it. Thus, *RQ2* receives a negative answer. A possible explanation is that OntoUML is more complex and participants had a very limited experience with it.

Regarding the User beliefs assessment, UML was more appreciated, probably due to the fact that OntoUML is more complex and was new to the participants who lacked any experience with the language. ITU opinions regarding OntoUML are strongly related to the results obtained for PEOU, because subjects will be reluctant to use a language whose learning barriers are higher than those of simpler alternatives. To answer *RQ3* considering the overall user beliefs, we can conclude that participants were hesitant to learn and use a novel modeling language, especially a complex one, in a short amount of time. However, the results indicate that performances, in terms of effectiveness, were better using OntoUML, although participants were not fully aware of this.

Previous OntoUML experience delivers better results. The two subjects with previous experience scored perfect results. Stronger opinions were revealed in the two subjects with previous experience. One had more negative beliefs; the other had better opinions than the average.

6   Conclusion

Conceptual modeling has been applied to complex domains, such as the human genome. In this paper, we describe an initial experiment for evaluating an 'ontological unpacking' method. The results showed that the participants' use of OntoUML, as needed for the ontological unpacking, achieved more correct responses than UML, although they took longer to respond. The experiment results revealed lesser intention to use, perceived ease of use, and perceived usefulness of OntoUML on the part of the participants. Based on these preliminary results, we plan to design broader experiments, that will include larger groups of participants, more heterogeneous subjects (in terms of age and background), and hypothesis testing based on the three research questions described here.

Acknowledgements.. This work was supported by ACIF/2021/117, INBIO2021/ AP2021-05, MICIN/AEI/10.13039/ 501100011033, and INNEST /2021/57 grants.

References

1. Almeida, J.P.A., Falbo, R.A., Guizzardi, G.: Events as entities in ontology-driven conceptual modeling. In: Laender, A.H.F., Pernici, B., Lim, E.-P., de Oliveira, J.P.M. (eds.) ER 2019. LNCS, vol. 11788, pp. 469–483. Springer, Cham (2019). https://doi.org/10.1007/978-3-030-33223-5_39
2. Bernasconi, A., Canakoglu, A., Ceri, S.: Exploiting conceptual modeling for searching genomic metadata: a quantitative and qualitative empirical study. In: Guizzardi, G., Gailly, F., Suzana Pitangueira Maciel, R. (eds.) ER 2019. LNCS, vol. 11787, pp. 83–94. Springer, Cham (2019). https://doi.org/10.1007/978-3-030-34146-6_8
3. García S., A., et al.: UML vs OntoUML analysis results [Data set] (2022). https://doi.org/10.5281/zenodo.6616114
4. García S., A., Guizzardi, G., Pastor, O., Storey, V.C., Bernasconi, A.: An ontological characterization of a conceptual model of the human genome. In: De Weerdt, J., Polyvyanyy, A. (eds.) CAiSE 2022. LNCS, pp. 27–35. Springer, Cham (2022). https://doi.org/10.1007/978-3-031-07481-3_4
5. García S., A., et al.: Towards the understanding of the human genome: a holistic conceptual modeling approach. IEEE Access **8**, 197111–197123 (2020)
6. García S., A., et al.: A conceptual model-based approach to improve the representation and management of omics data in precision medicine. IEEE Access **9**, 154071–154085 (2021)

7. Gray, T., De Vries, M.: Empirical evaluation of a new demo modelling tool that facilitates model transformations. In: Grossmann, G., Ram, S. (eds.) ER 2020. LNCS, vol. 12584, pp. 189–199. Springer, Cham (2020). https://doi.org/10.1007/978-3-030-65847-2_17

8. Guizzardi, G.: Ontological foundations for structural conceptual models. CTIT, Centre for Telematics and Information Technology (2005)

9. Guizzardi, G., Bernasconi, A., Pastor, O., Storey, V.C.: Ontological unpacking as explanation: the case of the viral conceptual model. In: Ghose, A., Horkoff, J., Silva Souza, V.E., Parsons, J., Evermann, J. (eds.) ER 2021. LNCS, vol. 13011, pp. 356–366. Springer, Cham (2021). https://doi.org/10.1007/978-3-030-89022-3_28

10. Guizzardi, G., et al.: Towards ontological foundations for conceptual modeling: the unified foundational ontology (UFO) story. Appl. Ontol. 10(3–4), 259–271 (2015)

11. Jošt, G., et al.: An empirical investigation of intuitive understandability of process diagrams. Comput. Standards Interfaces 48, 90–111 (2016)

12. Liaskos, S., Zhian, M., Jaouhar, I.: Experimental practices for measuring the intuitive comprehensibility of modeling constructs: an example design. In: Grossmann, G., Ram, S. (eds.) ER 2020. LNCS, vol. 12584, pp. 231–241. Springer, Cham (2020). https://doi.org/10.1007/978-3-030-65847-2_21

13. Moody, D.L., et al.: Evaluating the quality of information models: empirical testing of a conceptual model quality framework. In: Proceedings of the 25th International Conference on Software Engineering, pp. 295–305. IEEE (2003)

14. Pastor, O., Levin, A.M., Celma, M., Casamayor, J.C., Virrueta, A., Eraso, L.E.: Model-based engineering applied to the interpretation of the human genome. In: Kaschek, R., Delcambre, L. (eds.) The Evolution of Conceptual Modeling. LNCS, vol. 6520, pp. 306–330. Springer, Heidelberg (2011). https://doi.org/10.1007/978-3-642-17505-3_14

15. Reyes Román, J.F., Pastor, Ó., Casamayor, J.C., Valverde, F.: Applying conceptual modeling to better understand the human genome. In: Comyn-Wattiau, I., Tanaka, K., Song, I.-Y., Yamamoto, S., Saeki, M. (eds.) ER 2016. LNCS, vol. 9974, pp. 404–412. Springer, Cham (2016). https://doi.org/10.1007/978-3-319-46397-1_31

16. Verdonck, M., et al.: Comparing traditional conceptual modeling with ontology-driven conceptual modeling: an empirical study. Inf. Syst. 81, 92–103 (2019)

JUSMOD

The 1st International Workshop on Digital JUStice, Digital Law and Conceptual Modeling

Preface

Silvana Castano[1], Mattia Falduti[2], Cristine Griffo[2],
Stefano Montanelli[1,*], and Tiago Prince Sales[2]

[1] Department of Computer Science, Università degli Studi di Milano
{Silvana.castano, stefano.montanelli}@unimi.it
[2] Faculty of Computer Science, Free University of Bozen-Bolzano, Italy
{mattia.falduti, tiago.princesales}@unibz.it

Law plays a crucial role in almost every aspect of our life, both public and private. Thousands of legal documents are constantly produced by institutional bodies, such as Parliaments and Courts, which constitute a prominent source of information and knowledge for judges, lawyers, and other professionals involved in legal decision-making. To cope with growing volume, complexity, and articulation of legal documents as well as to foster digital justice and digital law, increasing effort is being devoted to digital transformation processes in the legal domain. Conceptual modeling plays a crucial role in this scenario, to formalize features and nature of terminology used in legal documents and promoting the development and the adoption of legal ontologies, shared vocabularies, and open linked data about legislation, case law, and other relevant legal information. Furthermore, advanced functionalities for legal data and process modeling and management are advocated, embracing modern technologies like Semantic Web, NLP, and AI, to enable semantic text search and exploration, legal knowledge extraction and formalization, legal decision-making, and legal analytics.

JUSMOD 2022, the *1st International Workshop on Digital JUStice, digital law and conceptual MODeling*, constitutes a meeting venue for a variety of researchers involved in digital justice and digital law crossing different disciplines besides computer science, like law, legal informatics, management, economics, and social sciences. The workshop represents an opportunity to share, discuss, and identify new approaches and solutions for modeling, analysis, formalization, and interpretation of legal data and related processes. The workshop program includes 5 accepted papers selected by the Program Committee after the peer-reviewed process and the invited keynote contribution on *Legal Knowledge Representation and Reasoning in the Semantic Web* by Enrico Francesconi (Legal Information Institute - Italian National Research Council).

Unsupervised Factor Extraction
from Pretrial Detention Decisions
by Italian and Brazilian Supreme Courts

Isabela Cristina Sabo[1](✉)(iD), Marco Billi[2](iD), Francesca Lagioia[2,3](iD),
Giovanni Sartor[2,3](iD), and Aires José Rover[1](iD)

[1] Department of Law, Federal University of Santa Catarina, Florianópolis, Brazil
isabelasabo@gmail.com
[2] CIRSFID - Alma AI, Alma Mater Studiorum- University of Bologna, Bologna, Italy
[3] Department of Law, European University Institute, Florence, Italy

Abstract. Pretrial detention is a debated and controversial measure
since it is an exception to the principle of the presumption of innocence.
To determine whether and to what extent legal systems make excessive use of pretrial detention, an empirical analysis of judicial practice
is needed. The paper presents some preliminary results of experimental
research aimed at identifying the relevant factors on the basis of which
Italian and Brazilian Supreme Courts impose the measure. To analyze
and extract the relevant predictive-features, we rely on unsupervised
learning approaches, in particular association and clustering methods.
As a result, we found common factors between the two legal systems in
terms of crime, location, grounds for appeal, and judge's reasoning.

Keywords: E-justice · Case factors extraction · Machine learning ·
Association rules · Clustering · Criminal law · Pretrial detention

1 Introduction

In criminal proceedings, pretrial detention is debated and controversial, since it
is an exception to the fundamental principle of the presumption of innocence, by
depriving defendants of their liberty at the initial stages of proceedings, before
their guilt is proven. The conditions under which such a measure is legitimate
include, for instance, the reasonable suspicion of the person having committed
the offence, the necessity to prevent defendants from absconding or committing
further offence(s), and the risk of interfering with the course of justice during
pending procedures. Their occurrence is subject to a case-by-case evaluation,
based on the judge's discretionary assessment. Moreover, the remand measure

This research has been supported by Brazilian Institutional Program for Internation-
alization (CAPES/PrInt); ADELE (Analytics for Decision of Legal Cases, EU Justice program Grant (2014–2020); COMPULAW (Computable law), ERC Advanced
Grant (2019–2024); LAILA (Legal Analytics for Italian Law), MIUR PRIN Programme
(2017).

R. Guizzardi and B. Neumayr (Eds.): ER 2022 Workshops, LNCS 13650, pp. 69–80, 2022.
https://doi.org/10.1007/978-3-031-22036-4_7

shall last no longer than necessary to achieve the objectives pursued by the law [7]. Unfortunately, while there have been numerous studies on the legal framework governing pretrial detention, limited research has been carried out to date into the practice of pretrial detention decision-making. In this regard, Italy and Brazil are interesting fields of investigation[1]. According to the World Prison Brief latest rates[2], in both countries, approximately 30% of the prison population are pretrial detainees. In this context, our research is aimed at identifying the relevant factors on the basis of which Italian and Brazilian Supreme Courts impose the pretrial detention –more exactly, maintain rather than reform, decisions on this matter by lower courts-, as well as how such factors relate to each other. To this end, we built two different corpora of Italian and Brazilian judicial decisions, as detailed in Sect. 2. Section 3 describes the unsupervised learning approaches, in particular association and clustering methods, used to analyse and extract the relevant predictive-features from the documents in the corpora. Section 4 reports the experimental setup and the results, as well as delineates commonalities and differences between the two legal systems. Section 5 concludes and outlines possible future research lines. This project follows recent attempts at explaining decision-making systems through factor-based reasoning, justifying decisions on the basis of legal features of a case [9,10]. In order to identify the legally relevant factors, described by [2] as case decision predictors, we followed recent experiments as seen in [5].

2   Datasets

We built two different datasets of Brazilian and Italian judicial decisions, as we could not find any existing data collections to help augment our own. The Brazilian corpus consists of 2,018 documents, collected from the official Brazilian Supreme Court's website (stf.jus.br). Documents are structured in the following sections: (a) heading (lawsuit metadata), (b) summary of the judgment, (c) case report (including the grounds of appeal), (d) reasons and decision of the judge-rapporteur, (e) votes of the other judges (when they differ from the judge-rapporteur), and (f) final decision. The Italian corpus consists of 718 judicial decisions by the Italian Supreme Court, downloaded from the DeJure database. Documents are structured according to the following sections: (a) heading (lawsuit metadata), (b) summary of the judgment, (c) case report (including the grounds of appeal), (d) reasons and (e) the final decision. In this regard, the main difference between the two corpora concerns the absence of dissenting statements in Italian rulings.

---

[1] For more information visit "Brazil has the world's 3rd largest prison population." https://www.conectas.org/en/noticias/brazil-worlds-3rd-largest-prison-population/ (2017), online; accessed 30 May 2022; and "A measure of last resort? The practice of pretrial detention decision-making in the EU." https://www.fairtrials.org/articles/publications/a-measure-of-last-resort-the-practice-of-pre-trial-detention-decision-making-in-the-eu/ (2016), online; accessed 30 May 2022.

[2] World Prison Brief. https://www.prisonstudies.org/, online; accessed 09 Jun 2022.

3   Methodology

In this section we briefly describe the general methodology and the unsupervised learning techniques we employed. We approach the research problem in two goals: (i) identification, aimed at extracting the relevant factors, and (ii) correlation, aimed at finding relationships between the extracted factors and judicial outcomes, i.e., whether Italian and Brazilian Supreme Courts maintain rather than reform decisions on pretrial detention. To this end, we adopted, for both the Brazilian and Italian corpora, a four-step process. First, we manually extracted some factors from judgments which we call *objective* factors, since they are clearly stated in the text. Second, we addressed the association task to find possible relationships between these *objective* factors and the decision outcomes. Third, to automatically extract further relevant features, we split each dataset into 2 subsets, on the basis of the outcome of the decisions. Finally, we applied clustering methods to each subset in order to detect what we name *subjective* factors, i.e., those that are more difficult to identify. Note that we did not apply association methods to the *subjective* factors, since the 2 corpora were already split depending on their outcome. To perform our experiments, we have relied on existing implementations and standard methods, including the open-source software Orange 3 [6] and Carrot2 [15], as detailed in Sect. 4. In Sects. 3.1 and 3.2, we briefly explain association and clustering methods.

3.1   Association

To identify relationships between factors and outcomes, we extracted association rules having the forms $x \rightarrow y$, where x is a set of factors and y is one of the two outcomes. For each rule, we determined its support and confidence, namely (a) the proportion of the cases in which both the antecedent x and outcome y are satisfied (the likelihood of finding x and y cases), as a fraction of all cases in the dataset, (b) the proportion of cases in which outcome y is satisfied, as a fraction of all cases satisfying factors x (the likelihood of x cases have outcome y).

$$s(x \rightarrow y) = \frac{Frequency(x,y)}{N} \ ; \ c(x \rightarrow y) = \frac{Frequency(x,y)}{Frequency(x)} \quad (1)$$

In particular, we applied the FP-Growth association algorithm to scan the whole data and find the rules which satisfy given support thresholds. Then the rules were represented as a conditional tree, which saves the costly dataset scans in the subsequent mining processes [8].

3.2   Clustering

Clustering is an unsupervised learning task used to uncover hidden patterns in unlabeled data [12]. Considering that documents may present common factors, we adopted the so called soft clustering approach, whereby documents can be assigned to one or more clusters. In particular, we applied Hierarchical Clustering, which builds tree structures, by merging documents, and clusters of them, depending on similarities [1]. To assess similarities we used the cosine measure

[4]. Once clusters have been generated, we ran the Latent Semantic Indexing (LSI) algorithm, which captures the underlying semantics of textual documents and computes how words relate to each other, so as to reveal the occurrences of topics within the corpora [16]. We also used the Lingo algorithm, which extracts frequent phrases from documents, under the assumption that such phrases provide informative human-readable descriptions of topics. Among the techniques on which Lingo relies, we employed the LSI, aimed at discovering any existing latent structures of diverse topics. Finally, Lingo matches the cluster description with the extracted topics and assigns each document to one or more clusters. To select the best label for each cluster, it uses a score measure, based on cosine similarity [14].

4   Experiments and Results

As explained in Sect. 3, we addressed our research questions as identification and correlation goals. In the following we detail the experimental uptake, we report the results and make some considerations.

4.1   Manually Extracted Information

Following the first step, we manually extracted 5 *objective* factors: the prisoner status, the name of the judge rapporteur, the crime category, the crime location and the judgment date. In the following, we detail each factor and the values it may assume depending on the data.

1. *Prisoner Status*, i.e., the situation of the accused after the appeal ruling. This factor may have two alternative values, i.e., *released* and *not released*. Cases in which the Court replaced pretrial detention with house arrest, were considered as released.
2. *Judge Rapporteur*, i.e., the judge who furnishes a report on the case at hand. The Italian data is characterised by a higher variance compared to the Brazilian one, due to the different number of seats in the two Supreme Criminal Courts: at least 35 in the Italian Supreme Court, regularly replaced[3], versus 11 seats in the Brazilian one, where judges have a permanent position.[4]
3. *Crime*, i.e., the general category to which the committed crime belongs to, under the Brazilian and Italian criminal laws. In particular, we identified four main categories: (i) "crimes against the person", (ii) "crimes against property", (iii) "drug-related crimes", and (iv) "criminal organization".
4. *Location*, i.e., the place where the crime took place. While in Brazil it corresponds to a state, in Italy it is represented by a regional capital.
5. *Date*, i.e., when the judgment was issued. It corresponds to the ruling year.

---

[3] Corte di Cassazione (Area Penale): https://www.cortedicassazione.it/corte-di- cassazione/it/area penale.page/, online; accessed 30 May 2022.

[4] Supremo Tribunal Federal: https://portal.stf.jus.br/ostf/, online; accessed 30 May 2022.

Following the second step, we run experiments by employing the FP-Growth association algorithm (see Sect. 3). Table 1 indicates the specific parameters we adopted. To generate a set of reliable rules having *Released* as a consequent, we had to lower the required support and confidence scores (given the smaller number of realise-cases being present in each dataset).

Table 1. Association setup parameters.

Technique	Tool	Consequent itemset	Parameters
FP-Growth	Orange 3	BR *Not released*	Min. Supp. 4%, Min. Conf. 70%
		IT *Not released*	Min. Supp. 4%, Min. Conf. 70%
		BR *Released*	Min. Supp. 1%, Min. Conf. 40%
		IT *Released*	Min. Supp. 1%, Min. Conf. 40%

Tables 2 and 3 show some selected results. In particular, we report the rules presenting a certain degree of similarity within the two corpora.

Table 2. Association rules in Italian dataset.

No	Antecedent	→	Consequent	Supp	Conf
1	Criminal organization, Reggio Calabria	→	Not released	6,6%	93,8%
2	Drug law crime	→	Not released	23,8%	84,0%
3	Napoli	→	Not released	14,5%	82,0%
4	2019	→	Not released	4,1%	96,8%
5	Crime against property, criminal organization	→	Not released	7,2%	82,5%
6	2013, drug law crime, Napoli	→	Released	1,1%	88,9%

Table 3. Association rules in Brazilian dataset.

No	Antecedent	→	Consequent	Supp	Conf
1	Judge rapporteur MA	→	Not released	39,8%	82,2%
2	Drug law crime	→	Not released	30,5%	73,6%
3	São Paulo	→	Not released	31,9%	73,9%
4	2019	→	Not released	22,7%	94,4%
5	Crime against property, criminal organization	→	Not released	4,1%	81,1%
6	2013, drug law crime, São Paulo	→	Released	1,0%	47,4%

As we can note from rules no. 2 and no. 5 within the Italian and Brazilian datasets, drug-related crimes as well as the combination of criminal organization and crimes against property, are factors usually related to the *not released* outcome. The same is true for the date factor 2019, the locations São Paulo and Naples, as shown in rules no. 3 and no. 4 in the two tables. Conversely, rule no. 6 in both datasets shows a relationship between the *released* outcome and the combination of date 2013, drug-related crimes and the location, respectively Naples and São Paulo. However, it should be noted that in the Brazilian dataset the confidence of this association rule is lower compared to the Italian one. From a general perspective, results show highly reliable association rules for the *not released* outcome within the two datasets. Conversely, we did not find association rules related to the *released* outcome with high confidence. This remains true even by reducing the confidence threshold.

4.2   Automatically Extracted Information

Following the third step, we split each corpus into two subsets, containing respectively the judgements for the defendant (*Released*) and for prosecution (*Not released*): in the Italian corpus, the first subset contains 614 judgements, and the second 104; in the Brazilian corpus respectively 1,503 and 515. We applied pre-processing techniques before clustering: normalization, tokenization combined with regular expressions, stemming, filtering of stop words and n-grams with $n = 2$ [12]. To encode sentences, in an effort to make our method as general as possible, we opted for well-established approaches. For the Lingo algorithm, we used the Bag of Words (BOW) model [11,17]. In this model, one feature is associated with each word in the vocabulary. The value of each feature is usually computed as the $TF - IDF$ score, and measures the importance of the corresponding word. For the Hierarchical algorithm, we used Word Embeddings, a popular technique for language models and deep learning applications [3,13]. The parameters adopted for clustering are reported in Table 4, depending on the outcomes and the number of documents in each subset.

Table 4. Clustering setup parameters.

Technique	Tool	Subset	Parameters
Lingo	Carrot2	IT *Not released* and *Released*	Cluster Count Base* 15%
		BR *Not released* and *Released*	Cluster Count Base 10%
Hierarchical clustering	Orange 3	BR and IT *Not released*	Height Ratio* 30%
		BR *Released*	Height Ratio 30%
		IT *Released*	Height Ratio 60%
LSI	Orange 3	All	3 Topics

*Measures used to calculate the number of clusters based on the number of documents on input.

Following the last step, for clustering, we rely on the Lingo algorithm, Hierarchical clustering and LSI. Tables 5, 6, 7 and 8 report some results obtained by using Lingo, sorted by highest score.

Table 5. Lingo clusters and labels in Italian *Not released* subset.

No.	Label and cluster	DN	Score	Type	Outcome
1	Maggio 2013 (C26)	61	36,15	Date	Not released
2	**Nullità dell'interrogatorio dell'indagato (C10)**	63	35,53	Grounds	Not released
3	**Termini di fase previsti dall'art 303 (C4)**	79	35,47	Grounds	Not released
4	Gravità indiziaria delle esigenze cautelari (C23)	61	33,05	Reason	Not released
5	Ipotesi di cui all'art 304 (C24)	61	32,22	Grounds	Not released
6	Napoli Emessa in data (C26)	61	31,43	Location	Not released
7	Principio della presuzione (C12)	63	30,27	Grounds	Not released
8	Reato Associativo Reati Fine (C5)	78	24,65	Crime	Not released

Table 6. Lingo clusters and labels in Brazilian *Not released* subset.

No.	Label and cluster	DN	Score	Type	Outcome
1	Vítima compareceu (C27)	150	25,87	Reason	Not released
2	**Excesso prazo custódia perdurar 5 meses (C13)**	152	24,65	Grounds	Not released
3	Senhora Ministra C. L. Presidente Exatamente (C3)	151	24,23	Judge	Not released
4	Prática crimes tráfico drogas porte (C25)	150	22,34	Crime	Not released
5	**Nulidade absoluta processo (C23)**	150	20,75	Grounds	Not released
6	Prevista art 44 Lei n 11343 (C24)	150	17,22	Reason	Not released
7	Dezembro 2014 (C12)	152	16,98	Date	Not released
8	Natureza droga apreendida cocaína (C28)	149	10,06	Reason	Not released

We classified the obtained labels as follows: (a) *grounds* of appeal (i.e. elements alleged by the defendant); (b) the *reasons* of the decision (elements indicated by the judges); (c) the type of committed *crime*; (c) the *location* of the lower court; (d) the *date* of the Supreme Court judgment; (e) and the name of the *judge rapporteur*. In analysing the results, we found some difficulties since multiple labels had similar meanings, and certain documents were included in more than one cluster. From the *Not released* subset of the Italian corpus we extracted grounds of appeal such as the nullity of the defendant's interrogation (label no. 2), the expiration of the pretrial detention term (label no. 3), and the violation of the presumption of innocence principle (label no. 7). Lingo also extracted labels referring to manually identified *objective* factors, e.g., the location (Naples, label no. 6), the date (May 2013, label no. 1) and the crime type (criminal organization, label no. 8). Among the requirements needed to apply the pretrial detention measure, the seriousness of the risks (label no. 4) is also related with maintaining the prison order. From the Brazilian *Not released* subset we extracted similar grounds of appeal, such as the expiration of the pretrial detention term (label no. 2) and the procedural nullity (label no. 5). As reasons for judgment, we listed the victim's appearance in court (label no. 1), the impossibility of converting the prison into an alternative measure in cases of drug-related crimes (label no. 6), also depending on the nature of the drug seized (cocaine, label no. 8). Here we also identified manually extracted labels, such as the date (December 2014, label no. 7), the crime (drug law crime) and the judge rapporteur (C. L., label no. 3).

Table 7. Lingo clusters and labels in Italian *Released* subset.

No.	Label and cluster	DN	Score	Type	Outcome
1	L'interrogatorio di garanzia ex art 294 (C5)	12	42,16	Reason	Released
2	Periodi di sospensione di cui all'art 304 (C2)	14	34,52	Reason	Released
3	**Sostituizione degli arresti domiciliari (C3)**	14	34,52	Grounds	Released
4	**Difensore alle ore (C11)**	9	29,59	Reason	Released
5	Febbraio 2009 (C6)	11	26,45	Date	Released
6	**Doppio dei termini previsti dall'art 303 (C9)**	10	26,03	Reason	Released
7	Caso di regressione (C8)	10	24,13	Reason	Released
8	Tribunale di Catanzaro (C12)	8	18,53	Location	Released

Table 8. Lingo clusters and labels in Brazilian *Released* subset.

No.	Label and cluster	DN	Score	Type	Outcome
1	Rio de Janeiro RJ (C2)	57	36.85	Location	Released
2	**Constrangimento ilegal decorrente excesso prazo (C5)**	52	36.69	Reason	Released
3	**Regime inicial aberto requer (C10)**	52	33.96	Grounds	Released
4	**Impte Defensoria Pública (C3)**	57	29.59	Reason	Released
5	Empresas investigadas (C17)	42	22.30	Reason	Released
6	Junho 2017 (C14)	50	20.06	Date	Released
7	Furto insignificante (C21)	9	18.83	Crime	Released
8	G. M. Segunda Turma Habeas Corpus 112 (C12)	51	15.25	Judge	Released

As regards the *Released* outcome, in the Italian subset we can note as related reasons the procedural nullity involving the defendant's hearing (label no. 1) as well as the suspension of the prison term-limit and its expiration (labels no. 2 and no. 6). These reasons can also be framed as grounds, as they were alleged by the defendant. We can further identify the following reasons: the issues concerning the defender (label no. 4), cases returned to the previous grade of judgement (label no. 7), and the replacing imprisonment with less restrictive measures (house-arrest, label 3). Once again, we verify factors regarding the date (February 2009, label no. 6) and the location (Catanzaro Court, label no. 8). We also found similarities in the Brazilian *Released* subset in terms of judgment reasons and grounds of appeal, such as the expiration of the prison term and unlawful constraint (label no. 2), less restrictive measures (label no. 3) and appeal proposed by the public defender (label no. 4). Cases related to investigated companies are also a factor that we classified as a reason (label no. 5). Other labels verified are when the situation involves an insignificant burglary (crime, label no. 7) and the judge-rapporteur (G. M., label no. 8).

Tables 9, 10, 11 and 12 show some selected results from Hierarchical and LSI.

Table 9. Hierarchical clusters and LSI topics in Italian *Not released* subset.

Topics and cluster	DN	Type	Outcome
(C16)			
1: p, art, 2020, comma, n, sospension, termini, d, p p, 2	11	Grounds/	Not released
2: art 304, 304, termini, p, sospension, comma, 304 p, p comma, p p, è		Date	
3: tribunal, 3, riesam, 304, art 304, periodo, art 309, 309, 309 p, sospension			
(C19)			
1: p, ti, art, sez, rv, p p, 3, 1, cautelar, comma	27	Reason/	Not released
2: r, co, cautelar, sentenza, cautelari, esigenz cautelari, esigenz, associazion, stupefacenti, dott		Crime	
3: presunzion, art 275, 3, 275, r, interrogatorio, 275 p, co, comma, comma 3			

Table 10. Hierarchical clusters and LSI topics in Brazilian *Not released* subset.

Topics and cluster	DN	Type	Outcome
(C12)			
1: hc, habea, art, corpu, habea corpu, ministro, min, tribun, prisão, voto	235	Crime	Not released
2: lei, art, pena, liberdad, tráfico, provisória, liberdad provisória, turma, crime, droga			
3: pena, provisória, liberdad, prisão, liberdad provisória, 33, art 33, regim, 4°, senhor			
(C21)			
1: crime, n°, lei, ministro, voto, tribun, habea, turma, marco, corpu	28	Crime	Not released
2: crime, código, criminosa, organização criminosa, lei, organização, s, art, sob código, código senha			
3: habea, habea corpu, corpu, crime, lavagem, n°, acórdão, relat, delito, dinheiro			

LSI returns green and red words, respectively indicating positive and negative weights. A positive weight indicates that a word is highly representative of a topic, while a negative weight indicates that a word is highly unrepresentative for that topic [6]. We tried to either lower or increase the number of topics with no real impact on the overall intelligibility of the results. Hence, the disadvantage of combining Hierarchical clustering and LSI is that we had to interpret single words rather than strings.

In the Italian *Not released* subset, we could identify factors already identified with Lingo, e.g., the suspension of the prison term and its expiration as grounds (C16 topics) on the one hand, and the seriousness of precautionary requirements, the connection between criminal organizations and drug-related crimes as a reason for applying pretrial detention (C19 topics) on the other hand. This factor can also be observed in the Brazilian *Not released* subset (C12 and 21 topics).

In the Italian *Released* outcome, we can observe similar results to those obtained with Lingo. In particular, we identified a few words referring to the hearing of the defendant, the general requirements for applying a precautionary

Table 11. Hierarchical clusters and LSI topics in Italian *Released* subset.

Topics and cluster	DN	Type	Outcome
(C7)			
1: p, art, p p, n, comma, cautelar, misura, 1, 2, ordinanza	54	Reason	Released
2: sentenza, appello, fase, interrogatorio, cort, misura, grado, p, pena, p p			
3: misura, 2, interrogatorio, bi, pena, comma, art 275, 275, comma 2, carcer			
(C4)			
1: p, art, comma, cautelar, custodia, n, 1, custodia cautelar, p p, termini	16	Reason	Released
2: art, termin, 1, comma, termini, p, fase, art 1, sentenza, durata			
3: misura, custodia, termin, p, sospens, 1, termini, giudic, custodia cautelar, sentenza			

Table 12. Hierarchical clusters and LSI topics in Brazilian *Released* subset.

Topics and cluster	DN	Type	Outcome
(C16)			
1: prisão, min, cautelar, hc, penal, liberdad, c., m., c. m., rel	20	Reason/	Released
2: direito, art, prazo, prisão, cautelar, rs, excesso, preventiva, prisão preventiva, duração		Judge	
3: pena, liberdad, lei, prazo, n°, privativa, sp, pena privativa, penal, privativa liberdad			
(C5)			
1: hc, min, prisão, turma, art, sp, habea, corpu, habea corpu, ministro	33	Reason/	Released
2: liberdad, turma, lei, art, m., c., c. m., dje, liberdad provisória, provisória		Judge	
3: primeira, primeira turma, g., m., g. m., prisão, domiciliar, min g., turma, prisão domiciliar			

measure (C7 topics), and the prison time expiration (C4 topics). In the Brazilian subset we identified a set of words referring to the time-limit of the prison (C16 topics), and house arrest as an alternative measure (C5 topics). Moreover, the algorithm extracted the name of two judges that are related to the release outcome(C16 and C5 topics).

5   Conclusion and Future Works

It is well known that the Brazilian and Italian Supreme Courts usually maintain, rather than reform, decisions on pretrial detention by lower courts. In our experiments, we aimed to go beyond this obvious observation and analyse the reasons behind such decisions. This may help us in determining whether this practice is legally correct or rather reflects the reluctance to overhaul decisions by lower courts. While our analysis does not provide a definitive answer, it shows a certain consistency in high court decisions. In both legal systems, clustering labels and topics point to factors in common, i.e., the excessive length of time spent in prison, and the time-limits established by the law are factors which support the release. On the other hand, crimes against property, drug-related crimes and involvement in criminal organizations are highly related to the maintenance of the pretrial detention measure. The same is true with regard to the locations of Naples and Sao Paulo, suggesting that in these places serious crimes are more recurrent. In the Brazilian dataset, we found relationships between the judicial outcome and the judge rapporteur. This situation is absent in the Italian dataset. This may be due to the higher variability of judges in this Court. Concerning the experimented methods, Lingo performs better than the Hierarchical clustering combined with LSI. Labels are immediately intelligible and contain meaningful information, from both computer science and a legal perspective. Moreover, the topics resulting from LSI could not be as easily linked to any relevant legal circumstance.

Future research includes structuring a dataset based on the factors highlighted and performing classification experiments through deep and classical machine learning to predict the outcome. In this sense, we also aim to obtain explanation of the predictions through the extracted factors.

References

1. Aggarwal, C.C.: Machine Learning for Text. Springer, Cham (2018). https://doi.org/10.1007/978-3-319-73531-3
2. Bex, F., Prakken, H.: On the relevance of algorithmic decision predictors for judicial decision making. In: Proceedings of the Eighteenth International Conference on Artificial Intelligence and Law, pp. 175–179 (2021)
3. Bojanowski, P., et al.: Enriching word vectors with subword information. Trans. Assoc. Comput. Linguist. **5**, 135–146 (2017)
4. Cichosz, P.: Data Mining Algorithms: Explained Using R. Wiley, Chichester (2015)
5. Dal Pont, T.R., et al.: Classification and association rules in Brazilian supreme court judgments on pre-trial detention. In: Kö, A., Francesconi, E., Kotsis, G., Tjoa, A.M., Khalil, I. (eds.) EGOVIS 2021. LNCS, vol. 12926, pp. 131–142. Springer, Cham (2021). https://doi.org/10.1007/978-3-030-86611-2_10
6. Demšar, J., et al.: Orange: data mining toolbox in python. J. Mach. Learn. Res. **14**(1), 2349–2353 (2013)
7. Duff, R.: Pre-trial detention and the presumption of innocence. Oxford University Press, Minnesota Legal Studies Research Paper 12-31 (2012)
8. Han, J., Pei, J., Yin, Y.: Mining frequent patterns without candidate generation. ACM SIGMOD Rec. **29**(2), 1–12 (2000)
9. Horty, J.: Reasoning with dimensions and magnitudes. Artif. Intell. Law **27**(3), 309–345 (2019). https://doi.org/10.1007/s10506-019-09245-0
10. Horty, J.F., Bench-Capon, T.J.: A factor-based definition of precedential constraint. Artif. intell. Law **20**(2), 181–214 (2012)
11. Hu, X., Liu, H.: Text analytics in social media. In: Aggarwal, C., Zhai, C. (eds.) Mining Text Data, pp. 385–414. Springer, New York (2012). https://doi.org/10.1007/978-1-4614-3223-4_12
12. Kotu, V., Deshpande, B.: Data Science, 2nd edn. Morgan Kaufmann (Elsevier Science), Cambridge (2019)
13. Mikolov, T., Chen, K., Corrado, G., Dean, J.: Efficient estimation of word representations in vector space. arXiv preprint arXiv:1301.3781 (2013)
14. Osiński, S., Stefanowski, J., Weiss, D.: Lingo: search results clustering algorithm based on singular value decomposition. In: Kłopotek, M.A., Wierzchoń, S.T., Trojanowski, K. (eds.) Intelligent Information Processing and Web Mining. Advances in Soft Computing, vol. 25, pp. 359–368. Springer, Heidelberg (2004). https://doi.org/10.1007/978-3-540-39985-8_37

15. Osiński, S., Weiss, D.: Carrot2 (2019)
16. Papadimitriou, C.H., et al.: Latent semantic indexing. J. Comput. Syst. Sci. **61**(2), 217–235 (2000)
17. Sebastiani, F.: Machine learning in automated text categorization. ACM Comput. Surv. (CSUR) **34**(1), 1–47 (2002)

Context-Aware Knowledge Extraction from Legal Documents Through Zero-Shot Classification

Alfio Ferrara[✉] [ID], Sergio Picascia[✉], and Davide Riva[✉]

Department of Computer Science, Universitá degli Studi di Milano,
via Celoria, 18, 20133 Milan, Italy
{alfio.ferrara,sergio.picascia,davide.riva}@unimi.it

Abstract. The extraction of conceptual and terminological knowledge from legal documents is a crucial task in the legal domain. In this paper we propose ASKE (Automated System for Knowledge Extraction), a system for the extraction of knowledge that exploits contextual embedding and zero-shot learning techniques in order to retrieve relevant conceptual and terminological knowledge from legal documents. Moreover, in the paper we discuss some preliminary experimental results on a real dataset consisting of a corpus of Illinois State Courts' decisions taken from the Caselaw Access Project (CAP).

Keywords: Legal knowledge extraction · Legal document retrieval · Zero-shot learning

1 Introduction

The extraction of knowledge from large textual corpora of legal documents is a crucial task for providing useful and relevant suggestions to legal actors, such as judges and lawyers, in handling new incoming cases. One of the main challenges in this context is that legal documents use a peculiar language and terminology, which makes it difficult to correctly classify and retrieve documents with unsupervised techniques based on standard information retrieval technologies. On the other hand, exploiting supervised learning techniques is also difficult due to the absence of sufficiently large annotated corpora.

In this paper we propose ASKE (Automated System for Knowledge Extraction), a system for the extraction of knowledge that exploits contextual embedding techniques to manage the meaning of legal concepts and terms, taking into account the context in which they are used; in particular, we employ natural language processing techniques for embedding chunks of documents [1], meaning that we map the plain text into a vector space of real numbers. ASKE operates cyclically and in a completely unsupervised environment, exploiting zero-shot learning (ZSL) techniques [4]; by ZSL we refer to the machine learning approach in which a classifier is required to predict labels for test data belonging to

R. Guizzardi and B. Neumayr (Eds.): ER 2022 Workshops, LNCS 13650, pp. 81–90, 2022.
https://doi.org/10.1007/978-3-031-22036-4_8

classes that were not observed during the training phase. At each cycle, ASKE classifies documents chunks according to a set of initial concepts, and retrieves relevant conceptual and terminological knowledge from them. This information will finally contribute to the enrichment of the ASKE Conceptual Graph (ACG).

In order to evaluate the ASKE performances, we discuss some preliminary experimental results on a real dataset consisting of a corpus of Illinois State Courts' decisions taken from the Caselaw Access Project (CAP). Finally, the paper presents an example of how ASKE can be exploited as a support tool for jurists, helping them in retrieving relevant legal precedents.

The paper is organized as follows. In Sect. 2, we present the ASKE methodology. In Sect. 3, we describe the conceptual model of the ASKE Conceptual Graph. In Sect. 4, we present some preliminary results obtained by evaluating ASKE on a corpus of real legal documents. In Sect. 5, we present the related work. In Sect. 6, we discuss our conclusing remarks.

2   The ASKE Methodology

Knowledge extraction in ASKE is performed through a sequence of operations that constitutes a cycle. Each cycle can be repeated for a predefined number of iterations, called generations. The extracted knowledge populates the ASKE Conceptual Graph (ACG), which is continuously updated at each cycle, allowing the model to incrementally classify documents, retrieve terminology and form new concepts, generation after generation. The four main phases of the ASKE cycle are shown in Fig. 1.

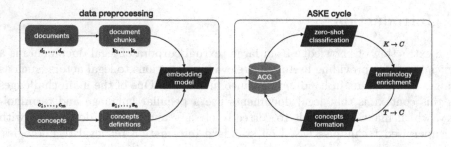

Fig. 1. Data preprocessing and ASKE cycle.

Data Preprocessing. The two main inputs of our model are a corpus of legal documents and a set of initial concepts. In particular, we consider a corpus of text documents $D = d_1, \ldots, d_n$, such as case law decisions or legal codes. Each document is preprocessed using a tokenizer that splits the text in a set of document chunks k and provides the lemmatized version of the terms t that constitute the documents. The set of initial concepts $C = c_1, \ldots, c_m$ can be provided by the user according to three different strategies: (i) the user provides just one or more terms t to use as initial concepts; (ii) the user provides one or more definitions s for each initial concept in form of a short text, such as a sentence; (iii) the user selects one or more concepts from an existing knowledge base. In case (ii) and (iii), each concept is already associated with one or more definitions s, given by the user in (ii) or provided by the knowledge base in (iii). In the first case, instead, we retrieve the definitions of each term associated to a concept from an external knowledge base by taking one definition s for each of the possible senses associated to a term. To the extent of this study, we gathered these definitions from WordNet, although our model is supposed to work with any external knowledge base provided. The last step of data preprocessing is to transform document chunks k and concept definitions s in their corresponding vector representation, projecting them in the same semantic space. This is achieved using Sentence-BERT [1], a modification of the original BERT model, which exploits siamese and triplets networks, being able to derive semantically meaningful sentence embeddings. The inputs for the embedding model will therefore be the set of document chunks K and the set of concept definitions S. Since a concept c_i may be associated to multiple definitions $s_{i1}, \ldots, s_{ij}$, we define its position in the embedding space as the centroid $\mathbf{c}_i$ of $\mathbf{s}_{i1}, \ldots \mathbf{s}_{ij}$. Document chunks and the set of initial concepts constitute together the first version of the ACG.

Zero-Shot Classification. Given the initial version of the ACG, ASKE is ready to perform the zero-shot multi label classification. In this phase, document chunks are assigned to none or multiple concepts, $f : K \to C$. A similarity measure σ, e.g. cosine similarity, between the embedding vector $\mathbf{k}$ of each document chunk and the embedding vector $\mathbf{c}$ of each concept is computed and, eventually, the two are associated if this similarity is higher than a predefined threshold α:

$$f(K, C) = \{c_i : \sigma(\mathbf{k}_j, \mathbf{c}_i) \geq \alpha\}$$

The hyperparameter α is crucial since it may remarkably affect the classification output: for example, choosing a high value of α will result in a high value of precision in the classification but potentially may find only a small set of document chunks for each concept.

Terminology Enrichment. For each concept c_i, ASKE retrieves the set of terms T_i appearing in the document chunks K_i associated to it. Then, these terms are placed in the embedding space computing the vector representation of their definition(s) s_{t_i} retrieved from an external knowledge base. In the case of polysemic terms, only the definition whose embedding is closest to the concept $\mathbf{c}_i$ is maintained. For each candidate terms the similarity σ between their embedding vector and the ones of the concept and of the document chunks is computed. The terms whose similarity is greater than the parameter β are taken into account and become candidates for enriching the terminology of the concept.

$$g(K, C, S_t) = \{t : \sigma(\mathbf{s}_{t_i}, \mathbf{c}_i) + \sigma(\mathbf{s}_{t_i}, \mathbf{k}_i) \geq \beta\}$$

The set of candidate terms is then sorted in descending order according to the similarity score. In addition, a learning rate γ can also be defined, representing the maximum number of terms which will be associated to a certain concept at each generation. Applying an upper bound γ and lower bound β ensures that, at each cycle, the process of terminology enrichment will include only a small set of terms that are supposed to be meaningful with respect to the concept at hand.

Concepts Formation. As a last phase, ASKE may introduce new concepts in the ACG by a process of concept formation. This process consists in applying, for each concept c_i, a clustering algorithm, such as affinity propagation [2], over the embedding vectors S_i of the terms t_i belonging to it. Among the resulting clusters, the one including the original definition of the concept c_i will continue to represent the original concept, preserving its existence in the ACG. For each remaining cluster, a new concept c_j will be generated, and its label will be represented by the term closest to the center of the cluster itself. Moreover, a derive from relation between c_i and c_j is defined in ACG.

3 The ASKE Conceptual Graph

The knowledge extracted during the ASKE execution cycles is stored in the ASKE Conceptual Graph (ACG), that is organized as shown in Fig. 2.

ACG is composed by two main entities, namely Concept and Term. A concept is associated with a label that represents its meaning in a synthetic and understandable way to a human user. This label is selected from the terms associated with the concept and corresponds to the term whose vector is closest to the mean of the vectors of all the terms associated with the concept. Concepts are also linked together by a relationship that describes how they were generated. In particular, the relationship derive from between two concepts c_i and c_j represents the fact that the concept c_j derives from the concept c_i, that is, it was formed starting from an aggregation of terms that were initially associated with c_i. As in all the ACG relations, also the relationship of derivation between two concepts is associated with the knowledge extraction cycle (generation) in which a concept was formed and the degree of similarity that associates a concept with the concepts derived from it. The concepts are also associated with the document chunks by the classification relationship. Note that ASKE performs a multi-label classification process, whereby a document chunk can be associated with multiple concepts. Moreover, a document chunk is associated via the part of relationship to a single document. In turn, each document can be associated to one or more document chunks. Finally, a concept consists of a collection of terms. The terms are represented by the term entity which is uniquely identified by the sense, that is a reference to a possible specific meaning of the term, and by the definition associated with that sense. The belongs to relationship stores the information about the set of terms associated with a concept in a certain cycle of execution of ASKE (generation) and with a certain degree of similarity (similarity). During the knowledge extraction process, each term acquired by ASKE is derived from a concept. This particular relationship is represented by the derive from relationship. Note that, even if the term is then associated with a different

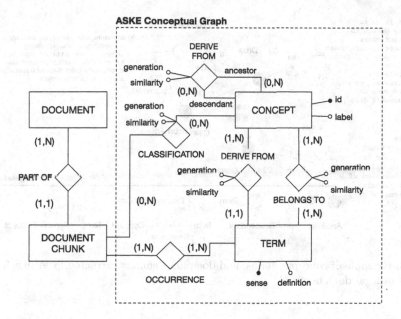

Fig. 2. Conceptual model of the ASKE conceptual graph

concept, the derive from relation still allows us to reconstruct the concept that led to the discovery of the term at hand. Terms also have a relationship occurrence with the document chunks. This relationship represents the presence of a term in one or more specific document chunk.

An example of how ACG is used to store the knowledge extracted from ASKE from a corpus of legal documents is shown in Fig. 3.

In the example, the concept C7, labeled Drug, was the initial concept. During the first generation (i.e., the first knowledge extraction cycle), C7 is associated, among the others, with the document chunk K1 with a similarity of 0.58 between the vector representation of K1 and the vector corresponding to C7. Through this association between C7 and the document chunks, new terms are extracted from the documents and associated with C7. In particular, a term associated with C7 is addition_0. Note that, in this example, where the term definitions are taken from WordNet, addition_0 corresponds to a particular synset of WordNet, whose definition is reported in the ACG. The knowledge extraction process is then repeated in subsequent generations, until, in the second generation, a group of terms initially associated with C7 and including addiction_0 is promoted to a new concept C14. One of the terms, namely drug abuse, is used to label the new concept and addiction_0 is now associated with it. However, we keep the derives from relation between addiction_0 and C7 to keep memory of the conceptual origin of addiction_0. In the subsequent generation 3, a new set of document chunks is associated with C7, including K2. This new document chunks lead to the discovery of new terms and to the generation of new concepts, including the new C24 (Crack) concept. Then, in the subsequent generation 4, C24 is also

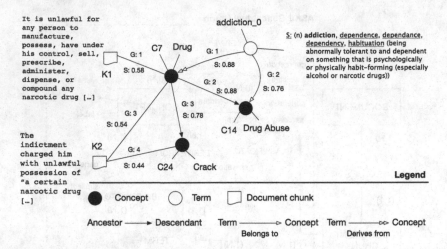

Fig. 3. Example of concepts, terms, and document chunks extracted by ASKE from a corpus of legal documents

associated with K2. Thus, K2 will be associated finally to more than one concept, namely, C7 (Drug) and, more specifically, C24 (Crack).

4   A Case Study on the Illinois Caselaw Corpus

Our aim here is to evaluate ASKE according to its ability to generate new terminological and conceptual knowledge starting from a small set of concepts and a possibly large corpus of documents. In the ASKE Conceptual Graph, we validate the concept derivation relationships, which link a concept to another that derived from it in any iteration, and the conceptual clusters of terms, i.e. the sets of terms that represent each concept, by submitting them to human evaluators.

Dataset. ASKE is applied to a subset of the Illinois Caselaw Corpus, a dataset from the CaseLaw Access Project[1]. Our dataset contains about 57,000 Case Law Decisions (CLDs) in the Illinois jurisdiction from 1771 to 2010. The CLDs were split into 9,5 million chunks, which constitute the input of ASKE together with a list of initial concepts, in the form of *(label; definition)* pairs. This procedure corresponds to the option (ii) described in Sect. 2, in which the definitions are provided by the users, here relying on the "Crime In Illinois 2020" report[2], issued by the Illinois State Police. The concepts correspond to 11 crime categories listed in the report, namely: homicide, rape, robbery, assault (also present with label battery), burglary, theft, vehicle theft, arson, human trafficking for commercial sex

---

[1] https://case.law.

[2] https://isp.illinois.gov/CrimeReporting/CrimeInIllinoisReports.

purposes (here labeled as prostitution) and human trafficking (for servitude purposes). Drug-related crimes, treated separately in the report, were added under the label drug, whose definition is taken from an American legal dictionary.[3]

Evaluation Process and Results. The evaluation process is twofold: in the ASKE Conceptual Graph, we want to i) validate concept derivation relationships and to ii) assess the coherence of conceptual clusters of terms. For an initial assessment, we relied on a restricted crowdsourcing campaign consisting of two classes of tasks and involving 8 workers.

4.1   Concept Derivation Relationships

For the evaluation of concept derivation relationships, we asked the workers to choose, in a set of 4 concepts given with their definition, the one that most closely relates to a concept in the question (i.e. *"Among the following, what is a concept that closely relates to concept* HOMICIDE *defined as* THE WILLFUL (NON-NEGLIGENT) KILLING OF ONE HUMAN BEING BY ANOTHER?*"*). There were 84 such tasks, of which 56 included a concept that ASKE derived directly from the one in the question, another one that was derived indirectly, and two other random concepts or *"None of the above"*. Other 28 out of 84 did not include the directly derived concept. Each task was executed by a minimum of one worker up to a maximum of 5. The workers choice is the ground truth of our evaluation, GA (Gold Answer). To the end of the evaluation, ASKE performs well on a task when the concept derived by ASKE from the concept in the question (either directly or indirectly) is included in the ground truth, i.e., the concept(s) chosen by the workers. This is formally represented by the evaluation function $e(t_i)$ that returns 1 if the concept retrieved by ASKE is in GA and 0 otherwise. It is easy to see that for ASKE the task was more difficult when there is a consensus among the workers, because the number GA of ground-truth answers is lower. To take into account this different value of the tasks, we computed a task score θ_i for each task as $\theta_i = 1 - (|GA_i|/4)$, where $|GA|$ is the size of the ground truth for the task i. Θ_T is the sum of θ_i for all the tasks T. The overall performance of ASKE over the set T of tasks is then evaluated as:

$$ASKE = \frac{1}{\Theta_T} \sum_{i=1}^{|T|} e(t_i)\theta_i$$

As a baseline, we computed the score of a hypothetical systems that randomly chooses one concept for each task. Since the probability of selecting one of the ground truth concepts is equal to $|GA|/4$, the performances of the baseline on a task i is evaluated as $(|GA|/4)\theta_i$. Moreover, we also computed the performances of ASKE by taking into account only the concepts directly derived from the concept in the task question (this last measure is called $ASKE_d$). The result of this evaluation is reported in Fig. 4.

---

[3] https://dictionary.law.com/Default.aspx?selected=343.

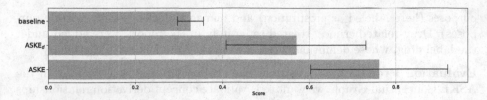

Fig. 4. Evaluation of ASKE concept derivation relationships

The evaluation was run on the Illinois Caselaw Corpus with the BERT model all-mpnet-base-v2[4] and 30 cycles. From this preliminary evaluation, we observe that ASKE may derive general or legal concepts by an accuracy of almost 80%. The $ASKE$ score is particularly affected by 11 tasks that received 4 out of 4 or 5 out of 5 answers not including the ASKE concept. Of such answers, 35 are related to the input concept Robbery, for which no worker identified legal concepts like Hearing, and Decree, nor generic concepts like Unreasonable as directly derived concepts.

4.2 Term Clusters

For term cluster evaluation, we asked the workers to assess if two terms, and their corresponding definitions in WordNet, were semantically related one to the other. In this evaluation, we have a correct result if two terms belong to the same concept in the ACG and they are considered semantically related by the workers (True Positive). If instead two terms belong to the same ACG concept but are not related according to the crowd we have a False Positive. Finally, we have a False Negative if two terms are considered similar but do not belong to the same concept in ACG. By exploiting this criterion, we computed the weighted Precision and Recall of ASKE according to different levels of worker consensus with the decision of ASKE, as shown in Fig. 5.

We note that in most of the tasks that have only one respondent the worker agrees with ASKE. The set of false positives is almost completely made of pairs of terms indicating components of the judicial process, the most common ones being: (action; trial), (trial; judgment), (trial; objection), (action; review). This suggests that ASKE is effective in grouping terms belonging to the judicial domain together, although the process of forming new concepts can be improved by increasing the system's ability to recognize sub-concepts within a more general concept.

5   Related Work

The main topic of interest for our work has been the zero-shot learning (ZSL) approach. ZSL is a problem setup in the field of machine learning, where a

---

[4] https://www.sbert.net/docs/pretrained_models.html.

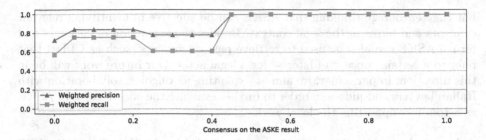

Fig. 5. Weighted precision and recall of ASKE according to the crowd consensus

classifier is required to predict labels of examples extracted from classes that were never observed in the training phase. It was firstly referred to as *dataless classification* in 2008 [3] and has quickly become a subject of interest, particularly in the field of natural language processing. The great advantage of this approach consists in the resulting classifier being able to operate efficiently in a partially or totally unlabeled environment.

It is possible to classify ZSL techniques according to three different criteria, as explained in [4]: the learning setting, the semantic space and the method. Firstly, ZSL can be applied on a completely unlabeled dataset, as in the original paper [3], or on a partially labeled one, like in [5]; with this last approach, called generalized ZSL, the goal of the classifier shifts to distinguishing between observation from already seen classes, and examples from unseen ones. Secondly, one may discern an engineered semantic space from a learned semantic space: the first is designed by humans and can be constructed upon a set of attributes [6] or a collection of keywords [7], while the second is built on top of the results of a machine learning model, as in the case of a text-embedding space [8]. Finally, ZSL methods can be divided in instance-based [9], whose focus is on obtaining examples for unseen classes, and classifier-based [10], which instead focus on directly building a classifier for unlabeled instances.

ASKE aims at classifying documents and extracting knowledge from them, building a conceptual graph on top of it. This process happens in a completely unsupervised environment, operating in a text-embedding space and applying an instance-based method; in particular, the employed method goes under the category of projection methods, which consist in labeling instances by collocating these examples in the same semantic space with class prototypes. It relies on a very limited initial knowledge and it is able to address the problem of term disambiguation.

6   Concluding Remarks

We presented ASKE (Automated System for Knowledge Extraction), a system for the extraction of knowledge that exploits contextual embedding and zero-shot learning techniques in order to retrieve relevant conceptual and terminological knowledge from legal documents. The experimental results are still preliminary, but they encourage thinking that ASKE can be effective in identifying relevant concepts and terms in the legal context. Besides the extraction of concepts and terms, ASKE can also be used to retrieve past documents (such as CLDs) that refer to a certain concept of interest for a legal actor. Our future work will be in this direction. In particular, we aim at exploiting an ongoing collaboration with Italian lawyers and judges in order to further evaluate the effectiveness of ASKE as a tool for supporting the legal work.

Acknowledgements. This paper is partially funded by the Next Generation UPP within the PON programme of the Italian Ministry of Justice.

References

1. Reimers, N., Gurevych, I.: Sentence-BERT: Sentence Embeddings Using Siamese BERT-Networks (2019)
2. Frey, B.J., Dueck, D.: Clustering by passing messages between data points. Science **315**(5814), 972–976 (2007)
3. Chang, M.-W., Ratinov, L., Roth, D., Srikumar, V.: Importance of semantic representation: dataless classification. In: Proceedings of the 23rd National Conference on Artificial Intelligence, AAAI 2008, vol. 2, pp. 830–835. AAAI Press (2008)
4. Wang, W., Zheng, V.W., Yu, H., Miao, C.: A survey of zero-shot learning: settings, methods, and applications. ACM Trans. Intell. Syst. Technol. **10**(2), 37, Article no. 13 (2019)
5. Xian, Y., Lampert, C.H., Schiele, B., Akata, Z.: Zero-Shot Learning - A Comprehensive Evaluation of the Good, the Bad and the Ugly (Version 4). arXiv (2017)
6. Lampert, C.H., Nickisch, H., Harmeling, S.: Learning to detect unseen object classes by between-class attribute transfer. In: IEEE Conference on Computer Vision and Pattern Recognition, pp. 951–958 (2009)
7. Qiao, R., Liu, L., Shen, C., van den Hengel, A.: Less is more: zero-shot learning from online textual documents with noise suppression (Version 1). arXiv (2016)
8. Xian, Y., Akata, Z., Sharma, G., Nguyen, Q., Hein, M., Schiele, B.: Latent Embeddings for Zero-shot Classification (Version 2). arXiv (2016)
9. Xu, X., Hospedales, T., Gong, S.: Transductive Zero-Shot Action Recognition by Word-Vector Embedding (Version 2). arXiv (2015)
10. Frome, A., et al.: Devise: a deep visual-semantic embedding model. In: Advances in Neural Information Processing Systems, vol. 26 (2013)

On the Lack of Legal Regulation in Conceptual Modeling

Kai von Lewinski[1] and Stefanie Scherzinger[2(✉)]

[1] Public Law, Media Law and Information Law, University of Passau,
Passau, Germany
kai.lewinski@uni-passau.de
[2] Scalable Database Systems, University of Passau, Passau, Germany
stefanie.scherzinger@uni-passau.de

Abstract. The act of conceptual modeling can be empowering, to the point where control over the conceptual schema becomes a *means of power*. In our discussion of conceptual modeling at the legal and computer science crossroads, we present real-world scenarios from the social domain where conceptual modeling has the power to actively or accidentally reshape the real world. For practical reasons, we primarily focus on examples from civil registers in Germany. We demonstrate that this power is de-facto unregulated by law (in Germany and elsewhere), and that the responsibilities for impactful decisions are rarely properly accounted for. Awareness seems to be—by and large—lacking, both in law and legal science, as well as within the modeling research community.

Keywords: Impact of conceptual modeling · History of conceptual modeling · Public registers · Database law

1 Introduction

In this article, we focus on conceptual modeling in the social domain where the conceptual schema not merely represents the domain, but often assumes an active role [20]: As a *means of power*, it might change the state of its domain [4, 13, 20], and can even be used to actively re-shape it, according to a political agenda or social stereotypes, or even just by accident. Aspects of this power inherent in the act of conceptual modeling have been subsumed in discussions on responsible software, such as the obvious problem of bias [5] and legal accountability [2].

However, the legal view on this power appears to be understudied. In this article, we assume this neglected legal point-of-view. Figure 1 shows an example from personal data in civil registers (such as the German "*Personenstandsregister*"), which record the civil status of persons (name, gender, martial status, etc.). The conceptual model is captured by an ER-diagram, which is then implemented as a

S. Scherzinger's contribution was funded by *Deutsche Forschungsgemeinschaft* (DFG, German Research Foundation) grant #385808805.

R. Guizzardi and B. Neumayr (Eds.): ER 2022 Workshops, LNCS 13650, pp. 91–101, 2022.
https://doi.org/10.1007/978-3-031-22036-4_9

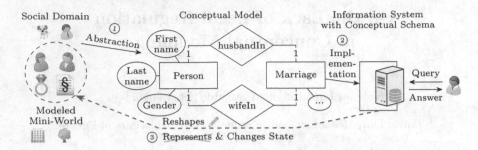

Fig. 1. ① A conceptual model abstracts from a mini-world and is ② implemented as a conceptual schema within an information system. ③ The conceptual schema impacts–or even *reshapes*–entities within the domain, possibly beyond the modeled mini-world.

conceptual schema within an information system (shown on the right side, illustration inspired by [13]). Users interacting with the system change the states of real-world entities, e.g., issuing a marriage license. The conceptual schema even has the power to enforce/protect monogamy by marriage, and thereby actively shapes the mini world (or even the entire social domain): a legal marriage severely limits the freedom to make a last will in favor of a lover.

Contributions. This article makes the following contributions:

- We provide historic and present-day examples that illustrate the power inherent in each stage of conceptual modeling in the social domain, focusing on *register data*. While our examples are set in Germany, we invite our readers to substitute matching scenarios from their own backgrounds.
- We pinpoint which stages in conceptual modeling are unregulated by law, and which indeed have legislative boundaries (such as data privacy and data protection). We show that there is a surprising lack of regulation.
- Our main contribution is to raise the awareness in the research communities of computer science as well as legal science, concerning both the extent of the power inherent in the act of conceptual modeling, as well as the apparent lack of regulation. We thus hope to inspire discussion and innovation.

Related Work. Overall, the legal perspective has been found to be underrepresented in conceptual modeling research [8]. Notions related to the act of conceptual modeling as a means of power have been expressed before, albeit in weaker terms: The responsibility of a model to *support action* (beyond promoting communication and understanding) is generally recognized [4,13,20]. Similarly, the idea of a model as an *instrument* has been called out [16].

Propositions to modeling concepts of legal power (not the act of power exercised in modeling) are present in the literature (e.g., [1,7]), yet this constitutes an orthogonal line of contributions.

Structure. In Sect. 2, we cover the historic and legal background. In Sect. 3, we discuss the stages in conceptual modeling, how power can be exerted, and to what extent this power is regulated. Sections 4 and 5, we discuss and conclude.

2   Background

Short History of Data Processing. Data processing (and with it the need for conceptual modeling) emerged alongside bureaucracies around 1750 B.C. in Babylonia, and around the same time in China: Life had become too complex (irrigation, food storage, living in cities, etc.) and exceeded the capacities of the human mind. This lead to the invention of lists and other administrative tools. This cultural technology of data processing was copied by the Ancient Romans. Based on lists and registers, entire empires could be governed. (In Europe, however, this information technology was forgotten during the Middle Ages, while being governed by means of the feudal *Personenverbandsstaat*[1]) [18].

Yet between 1500 and 1650, the modern state began to evolve in Europe, and bureaucratic data processing was (re-)invented [17]. The rising states and administrations used their growing informational power (censuses, registers, statistics) as a resource of power (over "outdated" feudal actors as well as over society in general). The states' bureaucratic powers grew vastly with the introduction of automated data processing (first punchcards, later computers), until today.

Data Protection Law and Other Data Regulation. First unease with a "datafying public administration" started to arise in the general public (cf. [14]) as well as among legal, computer science, and cybernetic scholars (cf. [12]) in the 1960s and 1970s. These growing concerns resulted in first privacy regulations (cf. the U.S. Privacy Act of 1974), particularly momentous in Europe and especially in Germany where the term *"Datenschutz"* ("data protection)" was coined around 1970 and where the first data protection laws originate from (Hessian State Data Protection Act of 1970; German Federal Data Protection Act of 1977).

Data protection laws regulate the processing of personal data as such but, by and large, do not address questions of databases or data processing infrastructure. This is particularly true for the EU but can also be observed in other legal systems around the world. Notably, data protection law has a specific microperspective, as its scope of application requires two preconditions to be met, covering only (1) the act of processing of (2) personal data of an individual. Consequently, conceptual modeling as well as designing and installing an information system is "allowed", but processing personal data (e.g., updating a single record) requires legal justification.

Larger-scale approaches towards database regulation have not been pursued, nor have they been prominently discussed by legislators or in academia (at least not to our knowledge). No general principles pertaining to public registers and other databases of public administrations exist.

(Only) in the EU such an approach was pursued with the Database Directive 96/6/EC which–perhaps surprisingly–does not address power exerted by

---

[1] The (German) *Personenverbandsstaat*, a technical term in historic sciences, describes a "state order based on personal ties" and refers to the feudal system of medieval Europe: Feudal ties formed people and rulers into a pyramid-style system order which helped manage large kingdoms, practically without any written documents.

databases but merely the intellectual property (IP) rights to the entirety of the database content (or database instance). Specifically, anti-trust law (cartel law) does not yet address databases of the public administration. However, in the early 2020s, legislative activities can be observed in the EU as to specific data regulation (esp. drafts for an "EU Data Act" and an "EU Digital Services Act").

Administrative Register Systems. In continental Europe, as well as in many other regions, systems for resident registration have been established during the last centuries and decades. Design and origin of these registration systems vary from country to country, yet the concept remains the same: the registration of every citizen and inhabitant for other or even all administrative purposes. In countries without such a compulsory comprehensive registration system, e.g., Anglo-Saxon countries, the administrative "full take" is usually performed by bureaucratic (sub-)systems that were originally designed for a different purpose, such as the Social Security Number in the U.S. Thus, comprehensively registering the entire population is a general phenomenon in the modern world.

3   Conceptual Modeling in the Social Domain

Figure 2 distinguishes generic stages in conceptual modeling (left-hand side of figure synthesized from illustrations in [9,15]). This chapter is structured as follows: In discussing each stage, we exemplify the power inherent in its execution, presenting specific, real-world scenarios. We then systematically discuss whether this power is regulated *de lege lata* (according to the law as it is) or whether it might even be regulated *de lege ferenda* (on the basis of new law).

3.1   Conceptualizing a Mini-World

In conceptual modeling, we identify a subset of the real world, which will then be modeled. Similarly, a given law also identifies and scopes a mini-world as a section of reality. Provisions as to "application" or "scope" exist in an explicit form in most legal acts; if they do not, implicit rules for their applicability exist.

Exertion of Power. Whereas *temporal scope* (date of promulgation) and *territorial scope* (the territory of a state) are often rather implicit, the *material scope* has to be made explicit in order to distinguish one law from the other. It is the law-maker who decides to whom the law applies. The following example illustrates how the scope of a law identifies a mini world.

Example 1. Civil register legislation applies to persons resident in a certain country (territorial scope) during the period of validity of the respective regulation (temporal scope), and with respect to certain persons. If you are not part of this "legal mini-world" you cannot have a full and decent legal status. This has consequences: Without such a status, you are not only excluded from social benefits and official identity documents, but you are also hindered to marry.

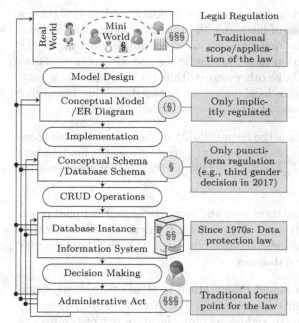

Fig. 2. Center: Stages (ovals) and artifacts (boxes) in conceptual modeling. Right: Legal regulation, from little ("§") to strongly regulated ("§§§"). Left: Arising legal problems (e.g., data privacy violations) trigger costly backtracking (see upwards arrows).

Regulation. In this early stage in conceptual modeling, the application of a law is strongly regulated.

Example 2. In Germany (on the federal level), legislation as to civil status registration dates back to the late 19th century, and covers all persons resident in Germany. You will not be subject to registration if you live outside the country, if you are a dog, a limited company, or an AI.

3.2 Conceptual Model Design

Modeling the mini-world produces a conceptual model, e.g., as formalized by an ER-diagram. Experience tells of the importance of this stage [13].

Exertion of power. The upcoming example focuses on modeling a person's name via attributes for the first- and last name.

Example 3. The first name does not come first everywhere: in some regions (Asia, Bavaria), the family name comes first and a person's individual name second. Some cultures and conventions do not fit the first name/last name scheme: Outliers are middle names (e.g., U.S.A.), father names (e.g., Spain), former aristocratic titles (e.g., Germany), maiden names, etc.

If a system is not able to correctly capture a person's name, it can be perceived as an act of violating their identity: The person is forced to use a false name in communication with administration (e.g., when filling in forms).

Regulation. Even seemingly basic modeling decisions can constitute acts of exerting power. We next examine how this power is regulated. In principle, "thoughts are free", and so are concepts and designs. But this freedom of thought (and concept and design) does not *per se* apply to the state and other public entities. In fact, the (modern Western) state does not enjoy *any* freedoms and rights, but rather bears competences and powers (which are subject to the law).

As a general doctrine, no general legal reservation (*legal proviso*) exists as to state actions (i.e., performed by public bodies); the state does not need an explicit legal permission for each and every thing it is doing (such as designing a conceptual model for the purpose of installing an information system). (Only) if an applicable law exists, the state under the law must comply (*primacy of law*).[2]

However, there is the restriction that, lawfully, state and administration may only act–informationally or in any other way–within their competences (and not *ultra vires*, i.e., beyond one's legal power or authority). This has direct implications as to which attributes of a person may actually be modeled.

Example 4. University education is the responsibility of the German states (*Länder*), not of the federal government. Consequently, student registration numbers (i.e., matriculation numbers) may not be issued by (and maybe not even recorded in) a federal database.

Example 5. Due to anti-discrimination laws and respective constitutional provisions, certain characteristics may not serve as relevant criteria for administrative decisions, such as race, religious beliefs, etc. Consequently, such attributes may not be stored in administrative databases.

In the examples above, the model design stage is at most *implicitly* regulated. Constitutional law systems (especially in the Western world) generally follow a liberal approach to human rights, keeping the state out of the lives of private persons (so-called "negative approach" or "1st generation human rights"). A more "positive approach" to human rights (so-called "2nd generation human rights") emphasizes the performance and service dimension of the state and its administration. While it is clear that a responsibility of a state is to provide security (as well as education, infrastructure, . . .), surprisingly (and to the best of our knowledge), law does not contain any specific obligation of the state to represent (parts of) the real world in an information system [15], even though administrations around the world are encouraged to improve in matters of digitization.

Overall, since not even a general regulation for conceptual modeling exists, there is no legal respective default rule. As long as no applicable legislation exists, the public administration is "free" (not) to design conceptual models.

3.3   Implementation of the Conceptual Model

Implementing the conceptual model, we obtain a conceptual schema, which may then be declared as a database schema.

Exertion of Power. Decisions made at this stage, such as fixing the exact attribute domains, are generally at a smaller-scale, but may have far-reaching consequences. In Germany, only a minimum subset of Unicode must be supported in administrative data exchange [6], so not all characters can be represented.

---

[2] Details go deep into constitutional theory and are specific to each constitutional system so that we cannot address this here in detail.

Example 6. It is part of the backstory of one of the worst terrorist attacks in Germany in recent years that the perpetrator, Anis Amri, could not be comprehensively monitored by the authorities because–besides other mishaps–his Arabic name had been entered in different spellings in Latin letters into the various police databases.

Example 7. German umlauts ("Schärzinger", "Löwinski") are a comparatively small technical challenge compared to the typographic richness of the Turkish or the Vietnamese (written) language, for example.

Problems with inconsistent transcriptions are also known to affect Russian names, and can be a setup for much frustration in communication with administrations when the same name is transcribed differently on different certificates.

The discriminatory potential of the declared attribute domain also becomes evident when we shift our focus from the name to the gender.

Example 8. Traditionally, German law recognized two genders, male and female. This enumeration type was extended in 2018 due to the introduction of a third gender as civil status, or the possibility to register a gender change. A person may now be as inter-sexual or as gender-fluid as he/she/it is, while in the hetero-normal binary gender order, he/she/it had to be either male or female— according to the principle *tertium non datur* (i.e., the *law of excluded middle*).

Ideally, such a change would propagate upwards in the modeling process, to the preceding stage: already in the design of the conceptual model, the attribute domain could be extended accordingly, and several modeling languages support enumeration types. Yet backtracking to an earlier step (indicated by the upwards arrows in Fig. 2) is often prohibitively expensive. Out of pragmatism, such changes are then directly carried out at the level of the database schema [10] (here synonymous to the conceptual schema). Upon changes to the database schema, the underlying conceptual models are often not co-maintained.

Regulation. The implementation of conceptual models is not regulated by law, neither in Germany nor (to our knowledge) elsewhere. Example 8 describes a rare exception, where court decisions demanded a specific implementation[3].

A further exception concerns data protection laws, which focus on the processing of individual records (and therefore discussed in more detail in the upcoming Sect. 3.4). Data protection law produces pre-effects, that already affect earlier stages in conceptual modeling. In some respects, data protection regulations like the EU GDPR request a specific implementation. Examples for this are security of processing (Article 32 GDPR) and Article 87 GDPR as to the specific provision for national identification numbers systems. As to databases, Article 25(1) GDPR demands "data protection by design" [3]. This means that action might have to be taken already at the currently discussed stage, or even

---

[3] BVerfG [German Federal Constitutional Court], decision of 10-Oct-2017, 1 BvR 2019/16—*Drittes Geschlecht* [Third Gender].

earlier, in model design (Sect. 3.2). Krasnashchok et al. [11] provide an overview on the related state-of-the-art in conceptual modeling research.

Beyond these selective regulations, no comprehensive legislation exists, not in Germany, not—despite the Database Directive 96/6/EC—in the EU, and—to our knowledge—not in any other legislations around the world.

Only very broad and general legal principles, such as purposiveness, exist. It asks for reasonable code, size, and format. This also includes the decision whether to use an existing conceptual schema, or whether to design from scratch.

3.4   Create-Read-Update-Delete (CRUD) Operations

Once the conceptual schema has been declared within an information system, create-read-update-delete operations (CRUD) modify the database state.

Exertion of Power. Things get "personal" (and therefore relevant under data protection law) when personal data is entered into a database.

Example 9. The first names "Sara" and "Israel" were quite popular and widespread within the Jewish communities in 20th century Europe. These names from the *tanach* demonstrated attachment to the history and religion of the Jewish people. This choice of names changed its character when it became obligatory for Jews in Nazi Germany to carry these names (as a means for identification as Jews and further discrimination).

Regulation. CRUD operations become legally relevant if and when data protection legislation is applicable (cf. Article 4 No. 1 GDPR). Although data processing (such as storing status data of an individual in a database) does not directly encroach the person's rights and interests, it is the conceptual idea of data protection law to shift protection upstream in the conceptual modeling process [19]. In consequence, a data controller has to adhere to a (rather large) set of rules, although no individuals have been impacted (yet).

3.5   Decision Making

Many state registers do not just register, but *manifest* a certain legal status. This status has more or less direct legal effects.

Example 10. While falling in (and out of) love may be a process, martial status captures a binary state, with far-reaching consequences such as to support obligations, limitations as to the last will, or legal consequences of infidelity.

Example 11. In many countries, the gender status in the civil register determines whether a person is required to do military service. As seen at the Ukraine borders, it can become a question of life and death (as some trans persons have reportedly experienced in the weeks of the Russian invasion in early 2022 when not allowed to leave the country because they we officially registered as "male").

According to the motto *quod non est in actis, non est in mundo* (what is not in the documents is not in the world), decision makers perform administrative acts based only on the database instance, not what actually holds in the real word. As this stage is the traditional focus point for the law, it is strongly regulated.

4   Discussion

Conceptual modeling is, to the largest part, a blind spot for the law. It might share the same starting point, as the provisions on its application define a *legal mini-world* (c.f. Sect. 3.1). Yet further stages of conceptual modeling are disregarded by the traditional concept of (Western liberal) law which focuses (only) on the immediate impact that a decision makes on a person (c.f. Sect. 3.5), and therefore the last stage in Fig. 2. Our observation is that indeed *every* stage in modeling has the potential to exert power. However, this has neither been addressed by legislators nor received broader discussion in (legal) academia. In fact, no comprehensive legal act exists as to databases and the power that the act of conceptual modeling holds.[4]

Apart from singular court decisions, the aspect of power by data ("data power"/"*Datenmacht*") is–so far–only addressed by data protection law (as in the GDPR). Its general regulatory approach is to regulate individual processing of an individual (natural) person's data by an individual processor. Thus, data protection law has a specific close-up perspective which focuses on a concrete encroachment but does not get, and even misses, "the big picture". At least, data protection law covers CRUD operations (c.f. Sect. 3.4).

The preceding stages have been addressed by law only in a very punctiform manner. Over time, law's blind spot has shrunk, but it persists, especially in the stages design and implementation (c.f. Sects. 3.2 and 3.3 respectively).

5   Conclusion

In this article, we have laid out how the act of conceptual modeling can constitute an act of exerting power, and that this power is largely unregulated by law. This raises the question whether this power should be actively regulated.

In general, law evolves at the rearguard of progress, relying on external triggers ("no traffic sign without a cause"). Yet informational power is a field that law has only been addressing since the 1970s (as illustrated in Fig. 2). 50 years constitute a large share in the history of modern computer science, but not from the perspective of legal science. Thus, any regulation of this power by law will require time, as well as interdisciplinary exchange.

---

[4] The EU Database Directive 96/6/EC, introduced in Sect. 2, does not meet this expectation: It establishes no more than an auxiliary database right, and only within the European Union. Furthermore, it merely covers the content of databases, not the conceptual schema as such. While originally envisioned as a universal model for IP law, it was not adopted by any other major legislations outside the EU.

Acknowledgments. We thank Meike Klettke and the anonymous reviewers for feedback on this article. We thank Thomas Kirz for expertly LATEXing the illustrations.

References

1. Bergholtz, M., Eriksson, O., Johannesson, P.: Towards a sociomaterial ontology. In: Franch, X., Soffer, P. (eds.) CAiSE 2013. LNBIP, vol. 148, pp. 341–348. Springer, Heidelberg (2013). https://doi.org/10.1007/978-3-642-38490-5_32
2. Cysneiros, L.M., do Prado Leite, J.C.S.: Non-functional requirements orienting the development of socially responsible software. In: Nurcan, S., Reinhartz-Berger, I., Soffer, P., Zdravkovic, J. (eds.) BPMDS/EMMSAD -2020. LNBIP, vol. 387, pp. 335–342. Springer, Cham (2020). https://doi.org/10.1007/978-3-030-49418-6_23
3. Danezis, G., et al.: Privacy and Data Protection by Design—From Policy to Engineering. European Network and Information Security Agency (2014)
4. Delcambre, L.M.L., Liddle, S.W., Pastor, O., Storey, V.C.: A reference framework for conceptual modeling. In: Trujillo, J.C., et al. (eds.) ER 2018. LNCS, vol. 11157, pp. 27–42. Springer, Cham (2018). https://doi.org/10.1007/978-3-030-00847-5_4
5. Friedman, B., Nissenbaum, H.: Bias in computer systems. ACM Trans. Inf. Syst. **14**(3), 330–347 (1996)
6. Gainsford, P.: Lateinische Zeichen in Unicode. Koordinierungsstelle für IT-Standards (KoSIT) Bremen (2012)
7. Griffo, C., Almeida, J.P.A., Guizzardi, G.: Conceptual modeling of legal relations. In: Trujillo, J.C., et al. (eds.) ER 2018. LNCS, vol. 11157, pp. 169–183. Springer, Cham (2018). https://doi.org/10.1007/978-3-030-00847-5_14
8. Härer, F., Fill, H.-G.: Past trends and future prospects in conceptual modeling—a bibliometric analysis. In: Dobbie, G., Frank, U., Kappel, G., Liddle, S.W., Mayr, H.C. (eds.) ER 2020. LNCS, vol. 12400, pp. 34–47. Springer, Cham (2020). https://doi.org/10.1007/978-3-030-62522-1_3
9. Kemper, A.: Datenbanksysteme—Eine Einführung, 10. Auflage. de Gruyter Oldenbourg (2015)
10. Klettke, M., Thalheim, B.: Evolution and migration of information systems. In: Embley, D., Thalheim, B. (eds.) Handbook of Conceptual Modeling, pp. 381–419. Springer, Heidelberg (2011). https://doi.org/10.1007/978-3-642-15865-0_12
11. Krasnashchok, K., Mustapha, M., Al Bassit, A., Skhiri, S.: Towards privacy policy conceptual modeling. In: Dobbie, G., Frank, U., Kappel, G., Liddle, S.W., Mayr, H.C. (eds.) ER 2020. LNCS, vol. 12400, pp. 429–438. Springer, Cham (2020). https://doi.org/10.1007/978-3-030-62522-1_32
12. Miller, A.: The Assault on Privacy. Penguin Group (USA) Incorporated (1972)
13. Olivé, A.: Conceptual Modeling of Information Systems. Springer, Heidelberg (2007)
14. Packard, V.: Naked Society. Pocket Books (1974)
15. Simsion, G.: Data Modeling Theory and Practice. Technics Publications (2007)
16. Thalheim, B.: Models are functioning in scenarios. In: Proceedings of the DAM-DID/RCDL, pp. 61–81 (2019)
17. Vismann, C.: Akten: Medientechnik und Recht. Fischer Taschenbücher Allgemeine Reihe (2001)
18. von Lewinski, K.: Die Geschichte des Datenschutzrechts von 1600 bis 1977. In: Freiheit—Sicherheit—Öffentlichkeit (Tagungsband der Assistententagung Öffentliches Recht 2008), Baden-Baden (2009)

19. von Lewinski, K.: Die Matrix des Datenschutzes—Besichtigung und Ordnung eines Begriffsfeldes. Mohr Siebeck, Tübingen (2014)
20. Wieringa, R.: Real-world semantics of conceptual models. In: Kaschek, R., Delcambre, L. (eds.) The Evolution of Conceptual Modeling. LNCS, vol. 6520, pp. 1–20. Springer, Heidelberg (2011). https://doi.org/10.1007/978-3-642-17505-3_1

Automated Extraction
and Representation of Citation Network:
A CJEU Case-Study

Galileo Sartor[1] , Piera Santin[2] , Davide Audrito[3(✉)] , Emilio Sulis[1] ,
and Luigi Di Caro[1]

[1] Computer Science Department, University of Torino, C. Svizzera 185, Turin, Italy
{galileo.sartor,emilio.sulis,luigi.dicaro}@unito.it
[2] CIRSFID - AI, University of Bologna, Via Galliera 3, Bologna, Italy
piera.santin2@unibo.it
[3] Legal Studies Department, University of Bologna, Via Zamboni 27, Bologna, Italy
davide.audrito2@unibo.it

Abstract. Although with some discrepancy, both in common law and in
civil law systems, previous judgments play an important role with respect
to future decisions. Traditional legal methodologies usually involve the
use of manual rather than automatic keyword search mechanisms to
retrace the steps of the judicial decision-making. However, these methods
are generally highly time-consuming and can be subject to different types
of biases. In this work, we present an automated extraction pipeline to
map and structure citations in rulings regarding fiscal state aids in the
case-law of the Court of Justice of the European Union. In particular,
by exploiting the available XML data in the EUR-Lex platform, we built
an end-to-end parser based on a set of regular expressions and heuristics, which is able to iteratively extract all citations, finally creating a
hierarchical structure of citations with their contextual information at
the paragraph level. Such data structure can be projected into a graphical representation, enabling useful visualization and exploration features
and insights, such as the diachronic study of the development of specific citations and legal principles over time. Our work suggests how the
exploitation and analysis of citation networks through automated means
can provide significant tools to support traditional legal methodologies.

Keywords: Legal knowledge extraction · Visual law · Citation
network · Digital justice

1 Introduction

According to the general theory of legal sources, in common law systems previous judgments are binding on future decisions, whereas in civil law systems
they only have a persuasive force. Over time this general distinction has been
attracting criticism. Some scholars argue, for instance, that there is no considerable discrepancy in the use of precedents in civil and common law systems

© The Author(s), under exclusive license to Springer Nature Switzerland AG 2022
R. Guizzardi and B. Neumayr (Eds.): ER 2022 Workshops, LNCS 13650, pp. 102–111, 2022.
https://doi.org/10.1007/978-3-031-22036-4_10

or that in both jurisdictions courts are bound to comply with precedents [6]. Moreover, the CJEU style of citation is quite peculiar, since it is inspired by a civil law model but it uses to invoke its own case law as common law courts do [13].

The importance of precedent citations in CJEU case law increased over the past years, and this appears also from the growing number of explicit citations in each judgment (e.g. the older one in our dataset does not have any citation, whereas the most recent one has nine direct citations with several recursive ones). For such reason, mapping precedents in CJEU case-law became crucial in order to better understand the development of the judicial framework.

Traditional legal methodology provides analytical tools to retrace the steps of judicial decision-making through citation networks. However, these methods are highly time-consuming and can be subject to bias, for instance due to non-empiric legal doctrine based exclusively on *opinio iuris*. In 2014, for example, Derlen and Lindholm [8] addressed the citation network of the Court of Justice of the European Union (CJEU) case-law. They found out that *Dassonville* and *Bosman* should be considered the most relevant landmark judgments, taking into consideration the frequency by which subsequent CJEU decisions cite these cases. On the contrary, handbooks use to attribute the greatest pioneering value to *Cassis de Dijon* and *Van Gend en Loos*. This example shows that computational analysis of law can have a significant role in supporting traditional legal research, in that citation networks can provide new insights in the understanding of judicial decision-making.

This work focuses on citation to previous rulings made by the CJEU in appeal decisions in the area of fiscal state aid. Such field is an interesting and significant case study, since it has a reduced legal basis and the discipline has been entirely developed through the CJEU judgments [22]. In particular, we choose to focus on appeals against decisions of the General Court (GC), because the relatively short time span (first judgment is of 1997) allows us to have a more homogeneous style of citations [21].

Furthermore, the Publications Office of the European Union developed an XML schema, which has been consistently applied to CJEU judgments. That allows the automatic extraction of references on the basis of affordable and original data, which guarantee certainty and precision of the results.

The main objective of our research is extracting and mapping the precedents cited in the CJEU case law in the area of fiscal state aid. The extracted data allows to offer a graphical representation that can be used to support legal studies, and may, in the future, enable further analysis and elaborations, for instance by way of NLP and other AI techniques [24]. The selected methodology is based on the annotated corpus of cases made available (in XML when possible) by the CJEU. This initial data is then further analyzed through the use of regular expressions to extract precise references to other CJEU cases. Regular expressions are used relying on heuristic rules, when the case is not yet available in XML form. This extracted data, saved in JSON and graph format, can then be analyzed through network analysis tools such as neo4j.

This experiment represents the first step for further comprehensive linguistic, legal and computer-science analysis.

Following this introduction, in Sect. 2 we briefly analyze the available works in the field of legal citation analysis. In Sect. 3 we outline the methodology proposed in our research. Section 4 then presents the main results achieved through the abovementioned methodology. Finally in Sect. 5 we sum up the initial results of our research, and outline possible future developments.

2   Related Work

Research in legal citation analysis has attracted interest over the last decades and is now becoming a well-established research area. The work of Fowler et al. on the case law citations by the Supreme Court of the United States [10] had a key role in identifying the benefits of citation analysis. The authors applied network analysis to address judicial citations, namely to explore the functioning of *stare decisis* and other issues related to the use of precedents. In the continental legal system similar approaches were followed to analyse the decisions of the Dutch Supreme Court [26], the CJEU [9] and the European Court of Human Rights (ECHR) [16].

Previous studies focused mainly on three directions: extraction of references, ranking of references and references labelling.

The automatic extraction of references has been developed mainly with regard to their different formats. To this end, Adedjouma et al. [1] used gazzetters and concept markers; Palmirani et al. [17] used regular expressions; Harasta [12] used Conditional Random Files (CRF); Leitner et al. [15] used BiLSTM neural networks. Other authors adopted methods beyond natural language processing, such as ML, deep neural networks and CRF to extract citations [2].

As shown in Langone [14], CRF and neural networks achieved similar performance in the extraction of citations, taking into account the number of references effectively identified in the legal sources at hand. As concerns other methods such as regular expressions, performances are dependent upon specific research objectives, e.g. whether authors aim to extract explicit or implicit references. Although effective, these experiments are not able to extract the entirety of references without gaps, differently from our methodology that affords to extract and map judicial citations contained in well-structured and annotated judgments published on CURIA, the official website of the CJEU.

Some studies are devoted to rank legal sources as a result of citation analysis based essentially on the number of codes' mutual citations [3].

In 2014, Derlén and Lindholm [8] argued that in-degree centrality is misleading when assessing the role of judicial decisions, because the relevance of a rarely cited judgment can increase if citations appear in important future decisions. Some domain-specific methods took into account other features, including the judicial instance [25].

Instead of ranking judgments, Sadl et al. [21] proposed to directly rank cited paragraphs to avoid confusion and inaccuracy, considering the peculiar structure of the CJEU system of references. For the same reason our extraction system allows to create an accurate, fine-grained, and reliable dataset (see Sect. 4).

In conclusion, multiple works address the task of identifying and labelling citations. To this end, several studies made use of citations' surrounding text [11], while others took into account different features, including signal words and patterns [7].

Network analysis in legislation found application in Sadeghian et al. [19], in which the authors focused on predicates, namely words that highlight the purpose or feature of citations without referring to the subject-matter at hand. They classified edges with 9 labels predicates were extracted through Conditional Random Fields and k-means classification with embeddings afforded to automatically classify the edges.

In 2018, Sadeghian automatically detected the purpose of cross references with a detailed semantic label set [20]. In a different way, Sulis et al. identified implicit citations by building a network of single portions of a text and undertaking binary classification tasks by way of co-occurrence network analysis [23].

While previous efforts in this field have been limited by the lack of well formed XML representations of case law, in our case the data made available by the European Union have a common structure for references in their XML schema. For this reason, the methodology presented in this paper may be easily extended to other legal domains of the European jurisdiction. In addition, our analysis allows a recursive regression that support the identification of pioneering interpretation statements, thus enabling a historical analysis of case law.

3   Methodology

In this research we focused on a subset of judgments, available on the EUR-Lex website, in the domain of fiscal state aid. The methodology described in this section is nonetheless expandable to other legal domains, and is only dependent on how citations are expressed in the original text of the decision.

In order to extract the desired information on the references to precedents contained in a judgment we started by downloading the XML representation available on the EUR-Lex platform[1].

The availability of this structured data is a step forward in the practical use of the information contained in the platform within research projects, thus simplifying, in our case, the initial assessment of the source text. The XML format is well documented [18] and includes different tags for citations. The one that we used is *REF.DOC.ECR*, a complete reference to previous cases. In particular, we only included references to a specific paragraph, assuming that they are the most legally-relevant. This could be easily changed in future extensions of the methodology and related tools.

It is worth mentioning that not all judgments are available in this format, and for those documents that are not in fact available we decided to use regular

---

[1] https://eur-lex.europa.eu.

expressions as a fallback. It is thus possible to extract information from both a well-formed XML representation, as well as the legal text, although parsing the XML file gives us information that has already been marked and verified.
In most cases the references in a CJEU case are of the form:

> [...] see, inter alia, judgments of 15 November 2011, Commission and Spain v Government of Gibraltar and United Kingdom, C-106/09 P and C-107/09 P, EU:C:2011:732, paragraph 73 [...]

This enables us to save a list of tuples, containing the cited case (the ECLI), and the cited paragraph number.

For each of the cited paragraphs, the relevant case is downloaded and the relevant portion of text is then extracted. On that section the reference extraction described above is carried out once again.

By evaluating citations in the single paragraph, we are able to collect only citations that may have a relation useful in the legal analysis. This can be seen in the outcome of our evaluation, where, for instance, the branches of the network containing the legal precedents are distinct from those about procedural rules. By following this procedure recursively we end up with a nested set of citations, with each citation having a list element, named *references*.

For the proper extraction of the paragraph number and its content, the parsing of the XML text is enhanced through the use of regular expressions. This functionality relies on the fact that the structure of these documents is generally quite uniform. Regular expressions have previously been adopted in NLP applications to extract structured information from legal sources/cases in plain text [17, 20].

The extracted data is then formatted in the standard JSON format, making it easily accessible to both non-technical persons and, again, to automated computer programs.

The code below shows an example of the information that is extracted for each reference, as it is saved in the JSON structure. Note that we store all different identifiers (ECLI, CELEX, and the Case Number) as well as the name of the judgment and text of the specific cited paragraph:

```
"ecli": "ECLI:EU:C:2011:732",
"text": "C-106/09 P and C-107/09 P Commission and Spain v Government of
    Gibraltar and United Kingdom [2011] ECR I-111137273",
"par_num": "NP0073",
"celex": "62009CJ0106",
"case_no": "C-106/09 P",
"xml_url": <url to the xml representation>,
"references": [...],
"par_text": "On the other hand, advantages resulting from a general
    measure applicable without distinction to all economic operators do
    not constitute State aid within the meaning of Article 87 EC (see, to
    that effect, Case C-156/98 Germany v Commission [2000] ECR I-6857,
    paragraph 22, and Joined Cases C-393/04 and C-41/05 Air Liquide
    Industries Belgium [2006] ECR I-5293, paragraph 32 and the case-law
    cited).",
```

On the basis of this information it is possible to use different tools to build visualization systems, citations' maps, and a database allowing further analysis of the citation network.

One such tool is neo4j, in which the above mentioned data can be imported and visualized through graphical or tabular representations of the citations. The system will, at a simple level, build a tree from the items it finds in the "references" sections of the database. Further analysis can be carried out on the basis of this representation, since it enables reasoning on the relations between the different nodes of the graph (cases and cited paragraphs). In the generated graph the cases are shown as yellow nodes, while the paragraphs are purple. There are two types of relations, the citation between two paragraphs (REFERSTO), and the case that a paragraph belongs to (BELONGSTO).

In the following section we shall see one of the ways in which we are able to query this database, and how it may assist in the legal analysis of precedents.

4   Results

By reasoning on the data structure defined in the previous section, the system is able to build a graphical representation of the relationships between citations. This representation can be useful to visualize the chronological development of citations and the evolution of legal principles.

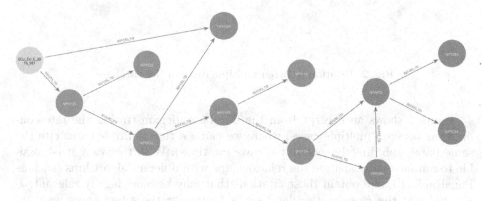

Fig. 1. Single branch of citations starting from a case.

In Fig. 1 we can follow a branch from the original case queried, in yellow, visualizing the sequence of paragraphs cited from previous decisions, in purple. All the cited paragraphs contain information on the original text, and the decision in which they are present. This can be used to analyse the evolution of a legal statement, from its origin in case law to its use as a consolidated principle [5].

This representation can be further enhanced with the ability to easily see the cited text as well as the relations between citations, thus assisting legal experts to assess the deep impact that the CJEU case law has on the European legal system.

It should be noted that the figure above shows only a single branch of the entire citation network. However, it is possible to expand the visualisation with different queries allowing to extract the needed case-specific information from the database. For instance, it is possible to view the relations between groups of judgments or to only expand the sections we are interested in.

With this information it will be possible, in future developments, to analyze the different citations, with complex algorithms to determine textual similarities and the relative importance of nodes in the network.

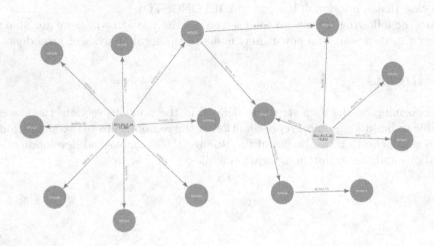

Fig. 2. Relation between multiple cases in the database

Figure 2 shows an excerpt from the database, demonstrating the interconnections between multiple cases, where we can see that multiple cases cite the same paragraph directly (not in recursive citations). With this view it is possible to automatically analyse the relationships with different algorithms (such as PageRank [4]), and obtain those citations that may be more legally relevant.

While at the moment the database is limited to the set of cases on state aid, it can be easily extended to a vast number of judgments, while maintaining the desired functionality. The main hurdle in applying this methodology to a broad set of cases is the consistency of the input data. From a short search on the EUR-Lex platform it is possible to see that cases are not always available in XML format, or where available, they are incomplete. This is not a blocking issue, as we have seen previously, since often the missing information, namely the paragraph number and the identifier, can be found in the text format and can be added after being parsed by the regular expressions.

A greater issue is where the citation is not identified with a unique code, such as ECLI or CELEX. In most cases it may be possible to find the judgment according to the parties involved in the proceeding, but especially in this legal area there may be several cases involving the same parties, such as the European Commission and a EU member State, e.g. Italy v. Commission. Heuristics may be useful to overcome this issue, for example by looking exclusively at earlier cases than the one at hand and supported by the fact that in more recent judgments citations are usually well-formed, at least in the legal text.

Furthermore, the sporadic unavailability of judgments translated in English must not be underestimated. However, it does not occur often, and it is fixed once the case is translated from the original language.

The system has been briefly tested on other legal fields and, at a first glance, it seems that judgments in fiscal state aid have a more consistent representation of the cited precedents, both in the XML format, and in the natural text. A possible reason for that is the absence of other legal instruments, such as directives or regulations. This fact makes the case law citations ever more important in order to determine the legal framework of fiscal state aid.

5   Conclusions and Future Work

The different representations made available by the system (as described in 4) may provide an interesting tool to support the analysis of case law in the European framework. The vertical representation (Fig. 1) can show branches of citations starting from a specific case and could be explored to better understand the context in which a decision was taken and its ability to affect future cases. This representation could also have practical uses for legal professionals by assisting the search and reasoning on legal principles and their evolution.

This initial work opens to a broad range of future research opportunities. The citation network could be enhanced through NLP and ML technologies, including by enriching the semantic comprehension of the text surrounding citations, possibly extracting meaningful representations and analyzing their role in judicial decision-making. The network could be expanded to show the nodes (i.e., paragraphs/principles) that are more frequently cited and to assess the evolution of nodes' citation frequency over time.

The full network can be used to better identify the most important paragraphs and can be adopted in legal research to compare the text between a citing and a cited paragraph. This relation could enhance the identification of new judicially-established legal principles.

On the legal side, the content of this network can be enriched by including more information extracted from legal cases. For instance, one could match citations to the outcome of the judgment, extract relevant keywords to map cases and their citations by the type of procedure, the specific matter, the originating state, or the legal effect.

Furthermore, the results obtained so far can be worked on with automated NLP solutions to extract more information from the cited paragraphs, without having to parse the whole text. This may be useful to determine whether cited paragraphs, originating from different judgments, are similar enough to share the intended meaning.

Another possibility is to link citations to the keywords extracted in all the available languages and possibly sorted through NLP tools and other automated means. This could be useful to determine how keywords are managed in the different European languages, as well as for the creation of a common, multilingual ontology for each relevant sector.

At first instance, this work relies on the information retrieved from the XML schema adopted by the CJEU itself. In so doing, the results obtained do not require any kind of interpretative bias. Moreover, the extraction is quite rapid and does not require any further manual annotation. Accordingly, it is possible to provide a tool that allows anyone to retrieve the citation network of a decision, just by providing its ECLI. Thus, the output will be immediately useful for the user, even in the absence of any further application of NLP technologies.

References

1. Adedjouma, M., Sabetzadeh, M., Briand, L.C.: Automated detection and resolution of legal cross references: approach and a study of luxembourg's legislation. In: 2014 IEEE 22nd International Requirements Engineering Conference (RE), pp. 63–72. IEEE (2014)
2. Bach, N.X., Thuy, N.T.T., Chien, D.B., Duy, T.K., Hien, T.M., Phuong, T.M.: Reference extraction from vietnamese legal documents. In: Proceedings of the Tenth International Symposium on Information and Communication Technology, pp. 486–493 (2019)
3. Boulet, R., Mazzega, P., Bourcier, D.: Network approach to the French system of legal codes part ii: the role of the weights in a network. Artif. Intell. Law **26**(1), 23–47 (2018). https://doi.org/10.1007/s10506-017-9204-y
4. Brin, S., Page, L.: The anatomy of a large-scale hypertextual web search engine. Comput. Netw. ISDN Syst. **30**(1–7), 107–117 (1998)
5. Brown, L.N., Kennedy, T.: The Court of Justice of the European Communities. Sweet & Maxwell, London (2000)
6. Civitarese, S.: A European convergence towards a stare decisis model. Rev. Digit. de Derecho Admin. **14**, 173 (2015)
7. De Maat, E., Winkels, R., Van Engers, T.: Automated detection of reference. In: Legal Knowledge and Information Systems: In: JURIX 2006 the 19th Annual Conference, vol. 152, p. 41. IOS Press (2006)
8. Derlén, M., Lindholm, J.: Goodbye van gend en loos, hello bosman? using network analysis to measure the importance of individual CJEU judgments. Eur. Law J. **20**(5), 667–687 (2014)
9. Derlén, M., Lindholm, J.: Is it good law? network analysis and the CJEU's internal market jurisprudence. J. Int. Econ. Law **20**(2), 257–277 (2017)
10. Fowler, J.H., Johnson, T.R., Spriggs, J.F., Jeon, S., Wahlbeck, P.J.: Network analysis and the law: measuring the legal importance of precedents at the U.S. supreme court. Polit. Anal. **15**(3), 324–346 (2007). https://doi.org/10.1093/pan/mpm011
11. Hamdaqa, M., Hamou-Lhadj, A.: An approach based on citation analysis to support effective handling of regulatory compliance. Future Gener. Comput. Syst. **27**(4), 395–410 (2011). https://doi.org/10.1016/j.future.2010.09.007
12. Harasta, J., Savelka, J.: Toward linking heterogenous references in czech court decisions to content. In: JURIX, pp. 177–182 (2017)

13. Koopmans, T.: Stare decisis in European law. Essays Eur. Law Integr, pp. 1957–1982 (1982)
14. Langone, D., Fulloni, A., Wonsever, D.: A citations network for legal decisions. In: New Frontiers in Artificial Intelligence. Lecture Notes in Computer Science. Springer International Publishing (2020)
15. Leitner, E., Rehm, G., Moreno-Schneider, J.: Fine-grained named entity recognition in legal documents. In: Acosta, M., Cudré-Mauroux, P., Maleshkova, M., Pellegrini, T., Sack, H., Sure-Vetter, Y. (eds.) SEMANTiCS 2019. LNCS, vol. 11702, pp. 272–287. Springer, Cham (2019). https://doi.org/10.1007/978-3-030-33220-4_20
16. Olsen, H.P., Küçüksu, A.: Finding hidden patterns in ecthr's case law: on how citation network analysis can improve our knowledge of ecthr's article 14 practice. Int. J. Discrimination Law **17**(1), 4–22 (2017)
17. Palmirani, M., Brighi, R., Massini, M.: Automated extraction of normative references in legal texts. In: Proceedings of the 9th International Conference on Artificial Intelligence and Law, pp. 105–106 (2003)
18. Publication Office, E.U.: https://op.europa.eu/en/web/eu-vocabularies/formex/
19. Sadeghian, A., Sundaram, L., Wang, D., Hamilton, W., Branting, K., Pfeifer, C.: Semantic edge labeling over legal citation graphs. In: Proceedings of the Workshop on Legal Text, Document, and Corpus Analytics (LTDCA-2016), pp. 70–75 (2016)
20. Sadeghian, A., Sundaram, L., Wang, D.Z., Hamilton, W.F., Branting, K., Pfeifer, C.: Automatic semantic edge labeling over legal citation graphs. Artif. Intell. Law **26**(2), 127–144 (2018). https://doi.org/10.1007/s10506-018-9217-1
21. Sadl, U., Tarissan, F.: The relevance of the network approach to European case law. reflexion and evidence. In: New Legal Approaches to Studying the Court of Justice (2020). https://hal.archives-ouvertes.fr/hal-03098351
22. Schön, W.: Tax legislation and the notion of fiscal aid: a review of 5 years of European jurisprudence. In: Richelle, I., Schön, W., Traversa, E. (eds.) State Aid Law and Business Taxation. MSTLPF, vol. 6, pp. 3–26. Springer, Heidelberg (2016). https://doi.org/10.1007/978-3-662-53055-9_1
23. Sulis, E., Humphreys, L., Vernero, F., Amantea, I.A., Audrito, D., Caro, L.D.: Exploiting co-occurrence networks for classification of implicit inter-relationships in legal texts. Inf. Syst. **106**, 101821 (2022)
24. Sulis, E., Humphreys, L.B., Audrito, D., Di Caro, L.: Exploiting textual similarity techniques in harmonization of laws. In: Bandini, S., Gasparini, F., Mascardi, V., Palmonari, M., Vizzari, G. (eds.) AIxIA 2021 - Advances in Artificial Intelligence, pp. 185–197. Springer International Publishing, Cham (2022)
25. Van Opijnen, M.: Citation analysis and beyond: in search of indicators measuring case law importance. In: JURIX, vol. 250, pp. 95–104 (2012)
26. Winkels, R., de Ruyter, J.: Survival of the fittest: network analysis of dutch supreme court cases. In: Palmirani, M., Pagallo, U., Casanovas, P., Sartor, G. (eds.) AICOL 2011. LNCS (LNAI), vol. 7639, pp. 106–115. Springer, Heidelberg (2012). https://doi.org/10.1007/978-3-642-35731-2_7

A Rule 74 for Italian Judges and Lawyers

Giulia Pinotti[✉], Amedeo Santosuosso, and Federica Fazio

European Center for Law and New Technologies, University of Pavia, Pavia, Italy
giulia.pinotti1991@gmail.com

Abstract. Rule 74 of the Rules of the Court (March 2022) states that any judgment of the European Court of Human Rights must contain some basic information (such as the names of the parties, agents, lawyers or advisers of the parties and a report of the procedure followed) followed by some well-defined steps of the decision: the facts of the case; a summary of the arguments of the parties; the legal reasons; operational provisions. Similar criteria are also established for the drafting of the appeals of lawyers.

This set of rules has the function of giving a rigid and predetermined structure to the introductory acts of the parties and to the decision of the Court with reference to both the purely formal elements and the essential elements of the judgement in a substantive sense.

The proposal for a Rule 74 for Italian judges is not an end in itself. The paper discusses how this, combined with a similar rule for lawyers' acts, can play an essential role in the application of AI technologies to law. Indeed, the availability of a dataset composed of decisions having very homogeneous structures can facilitate, as will be shown, the application of Natural Language Processing and machine learning techniques and finally a better performance in different activities such as prediction, summarization, knowledge extraction.

Keywords: Artificial intelligence · Judges · Legal analytics

1 Introduction

Conducting experiments with AI tools in a given domain – legal, in our case – requires not only a large amount of data (as is now well known) but also that these be of a defined quality.

Rule 74 of the Rules of the Court (March 2022) states that any judgment of the European Court of Human Rights must contain some basic information (such as the names of the parties, agents, lawyers or advisers of the parties and a report of the procedure followed) followed by some well-defined steps of the decision: the facts of the case; a summary of the arguments of the parties; the legal reasons; operational provisions. Similar criteria are also established for the drafting of the appeals of lawyers.

This set of rules has the function of giving a rigid and predetermined structure to the introductory acts of the parties and to the decision of the Court with reference to both the purely formal elements and the essential elements of the judgement in a substantive sense.

R. Guizzardi and B. Neumayr (Eds.): ER 2022 Workshops, LNCS 13650, pp. 112–121, 2022.
https://doi.org/10.1007/978-3-031-22036-4_11

The proposal for a Rule 74 for Italian judges is not an end in itself. The paper discusses how this, combined with a similar rule for lawyers' acts, can play an essential role in the application of AI technologies to law. Indeed, the availability of a dataset composed of decisions having very homogeneous structures can facilitate, as will be shown, the application of Natural Language Processing and machine learning techniques and finally a better performance in different activities such as prediction, summarization, knowledge extraction.

2 Methodologics for the Quantitative Analysis of a Legal Text: The Advantages of a Text-Based Approach

AI techniques for conducting legal analysis trials can be classified in various ways: according to their purpose, to the technology used, or to the role played by the object, in this case the legal text[1].

If we look at the most significant studies with a view to the approach used towards the text we can classified them into three main categories according to their research strategies: text-based approaches, metadata-based approaches, and mixed approaches. Text-based approaches rely only on the content of the initial data (in our case, the body of a decision) to build a structured description of the text that enables the algorithm to predict the outcome of the decision. Metadata-based approaches use background and context information, which can be inferred from the case, but do not rely directly on the text of the decision[2]. Mixed approaches, clearly, combine the other two approaches. The choice of one approach over another may be based on several legal, technical, or practical reasons[3].

From a practical point of view, the choice of the approach can depend on the available data: The existence of a database containing decisions whose metadata have already been tagged, for example, can make a metadata-based approach more immediate and less time-consuming; on the other hand, it is evident the availability of a dataset consisting of very structurally homogeneous decisions encourages a text-based approach. In fact, the more homogeneous the structure of the text, the more reliable the tagging can be.

The study published in 2016 by Aletras et al. is very consistent with what we have remarked concerning the relevance in the text-based approach of the homogeneous structure of the decision.[4] The study, as will be better seen in the next section, focuses on European Court of Human Rights decisions and has the aim to "build predictive models

---

[1] See A. Santosuosso – G.Sartor, La giustizia predittiva: una visione realistica, in Giurisprudenza italiana, 7/2022, in print.

[2] Katz, D.M.; Bommarito, M.J., II; Blackman, J. A general approach for predicting the behavior of the Supreme Court of the United States. PLoS ONE 2017.

[3] Santosuosso, A.; Pinotti, G. Bottleneck or Crossroad? Problems of Legal Sources Annotation and Some Theoretical Thoughts. Stats 2020, 3, 376–395. https://doi.org/10.3390/stats3030024.

[4] Aletras, N.; Tsarapatsanis, D.; Preo,tiuc-Pietro, D.; Lampos, V. Predicting judicial decisions of the European Court of Human Rights: A Natural Language Processing perspective. PeerJ Comput. Sci. 2016, 2, e93. See also the more recently published Medvedeva, M., Vols, M. & Wieling, M. Using machine learning to predict decisions of the European Court of Human Rights. *Artif Intell Law* 28, 237–266 (2020). https://doi.org/10.1007/s10506-019-09255-y.

that can be used to unveil patterns driving judicial decisions". The authors focus on the automatic analysis of cases of the European Court of Human Rights: The task is "to predict whether a particular Article of the Convention has been violated, given textual evidence extracted from a case, which comprises of specific parts pertaining to the facts, the relevant applicable law and the arguments presented by the parties involved".

In order to do so, they "formulate a binary classification task where the input of classifiers is the textual content extracted from a case and the target output is the actual judgment as to whether there has been a violation of an article of the convention of human rights". They create a dataset consisting of decisions on Articles 3 (Prohibition of torture), 6 (Right to a fair trial), and 8 (Right to respect for private and family life) of the Convention. They decide to focus on these three articles for two main reasons: "First, these articles provided the most data we could automatically scrape. Second, it is of crucial importance that there should be a sufficient number of cases available, in order to test the models. Cases from the selected articles fulfilled both criteria" (p. 7). At the beginning of the paper, they observe that their "main premise is that published judgments can be used to test the possibility of a text-based analysis for ex ante predictions of outcomes on the assumption that there is enough similarity between (at least) certain chunks of the text of published judgments and applications lodged with the Court and/or briefs submitted by parties with respect to pending cases". This supposed similarity is due to the structure of ECHR decisions, as will be discussed in the next paragraph.

The authors then explain the (technical) reason why the structure is so relevant: assuming that semantically similar words appear in similar contexts (where by contexts they mean specific parts of the decision), the authors find the group of words that best summarizes the content (and outcome) of a decision. As they stress, "we create topics for each article by clustering together N-grams that are semantically similar by leveraging the distributional hypothesis suggesting that similar words appear in similar contexts". The performance of the model is measured on the basis of the correctness of the prediction (since the result, the court's decision, is already known for the test cases). The average accuracy of the model for all cases analyzed is 79%. What you can see when looking at the results is that the most important feature is constituted by "circumstances", especially when combined with information from the topics. In the case of a violation of Article 6 of the ECHR, combining these two data gives an accuracy of 84%.

In the same direction also seems to go an interesting study published in 2019 by Zhong et al. The paper reports on an experiment in automatic extractive summarization of legal cases concerning post-traumatic stress disorder (PTSD) from the US Board of Veterans' Appeals (BVA)[5].

They "randomly sampled a dataset of single-issue PTSD decisions for this experiment. It comprised 112 cases where a veteran appealed a rejected initial claim for disability compensation to the BVA" They build a system able to summarize decision-relevant aspects of a given BVA decision by selecting sentences that are predictive of the case's outcome and they evaluate the system using a lexical overlap metric to compare

[5] Zhong, L.; Zhong, Z.; Zhao, Z.; Wang, S.; Ashley, K.; Grabmair, M. Automatic Summarization of Legal Decisions using Iterative Masking of Predictive Sentences. In Proceedings of the Seventeenth International Conference on Artificial Intelligence and Law, Association for Computing Machinery, New York, NY, USA, 17–21 June 2019; pp. 163–172.

the generated summaries along with expert-extracted summaries (i.e., selected sentences in the text, which summarize the decision) and expert-drafted summaries. The result of the study is evaluated (as already observed in line 178) using Rouge metrics 1 and 2, which measure the overlapping of words and word pairs. The authors compare (using Rouge 1 and 2) the results of manually extracted, drafted, and automatically extracted summaries. Comparing the work done by individual annotators (responsible for manually extracting the summaries) one obtains a very high score (greater than 0.8). The results of the comparison of the automatically extracted summaries with those extracted manually gives a result of about 0.65. Comparing the latter to sentences extracted randomly from decisions, one obtains almost identical results (around 0.63). This raises doubts about the metric chosen. The authors do not limit themselves to a quantitative analysis but also make a qualitative one, analyzing also the presence for each summary of a minimum set of information considered indispensable. This happens in about 50% of cases of automatically generated summaries.

This study, while not dealing with prediction, shows how even in this case the homogeneity of the texts was one of the reasons that led to the adoption of certain technological choices, specifically the use of text-based approaches. As they state, the "research goal was to discover if outcome-predictiveness in the dataset can serve as a proxy for such domain-model-like information. After examining the results, the answer to these questions is a qualified 'no', but some insight has been gained". Independently of the (self-evaluated) negative outcome of the research, it is interesting to note that even in this study the choice of the adopted technique is due to the thematic focus, structural homogeneity, and size of the dataset, which are expected to support research on automatically summarizing legal cases.

3 Data Homogeneity: From Tagging to Predetermined Decision Structure

A text in a very homogeneous dataset with a rather rigid structure more easily allows a text-based approach. In the analysis of case law, these characteristics should be specific to each decision. This apparently neutral consideration is very problematic within a legal system. The heterogeneity of decisions is due, in fact, not only to the intrinsic heterogeneity of the subject but also to the fact that there is a plurality of courts in each system, distributed by level and matter. In addition, even among courts (distributed over the territory) that have the same competence in terms of the matter, it might be hard to impose a rigid structure, in formal or non-formal terms, on decisions. The attempt to impose a subsequent structure by manual tagging meets the repeatedly mentioned problem of a bottleneck.

On a more general level, therefore, the inability to modify the data available (the body of each decision) might shift the focus on the creation of a database that collects not only information that can be easily extracted manually from the body of each decision (e.g., the court that pronounced it) but also context information (e.g., the party of a bill proposer). Needlessly to say, in this case, the problem of the bottleneck recurs, since not all the information (and especially that related to the context) can be extracted automatically, and it is necessary to introduce a phase in which texts are manually tagged by an expert in the field.

Tagging therefore proves to be an essential ally when the source dataset lacks the necessary homogeneity of structure. If we look, however, to the future in the use of AI tools in the legal sector, then we may wonder whether we cannot make choices right from the genesis of the text (the judgment) that may prove useful in the subsequent quantitative analysis. In the following paragraphs we will try to explain how the predetermination of formal and substantive elements in a judgment, as in the case of Rule 74, can be useful both for legal reasons (see par. 4) and, as just illustrated, for technical reasons, also for the correct implementation of a document builder (see par. 7).

4   What is the Structure of Rule 74?

We have tried to explain why rigid structure and homogeneity of the text are so relevant from a technical point of view. Now we discuss, from a legal point of view, how a judgment with these characteristics can be constructed, starting from the model developed by the European court of human rights.

The structure of ECHR decisions is determined by Rule 74 of Rules of the Court, which, after the basic information about the parties, the judges and the background, requires a clear description of (e) *an account of the procedure followed; (f) the facts of the case; (g) a summary of the submissions of the parties; (h) the reasons in point of law; (i) the operative provisions; (j) the decision, if any, in respect of costs*[6].

---

[6] "1. A judgment [...] shall contain (a) the names of the President and the other judges constituting the Chamber or the Committee concerned, and the name of the Registrar or the Deputy Registrar; (b) the dates on which it was adopted and delivered; (c) a description of the parties; (d) the names of the Agents, advocates or advisers of the parties; (e) an account of the procedure followed; (f) the facts of the case; (g) a summary of the submissions of the parties; (h) the reasons in point of law; (i) the operative provisions; (j) the decision, if any, in respect of costs; (k) the number of judges constituting the majority; (l) where appropriate, a statement as to which text is authentic. 2. Any judge who has taken part in the consideration of the case by a Chamber or by the Grand Chamber shall be entitled to annex to the judgment either a separate opinion, concurring with or dissenting from that judgment, or a bare statement of dissent".

Similar requirements are determined for applications to be made on the application form provided by the Registry[7]: "(e) a concise and legible statement of the facts; (f) a concise and legible statement of the alleged violation(s) of the Convention and the relevant arguments; and (g) a concise and legible statement confirming the applicant's compliance with the admissibility criteria laid down in Article 35 § 1 of the Convention."

The information parties and judges are requested to give can be divided into two main categories. The first regards the information about parties, applicants, representatives, lawyers, judges, addresses, telephone, fax numbers, e-mail and more: they constitute a sort of ID of the case. The second is more conceptual and regards the factual and legal substance of the case.

In Table 1 it is shown the correspondence between the parts of the judgement and those of the application. To be noted that the (e) part of the judgement and the (e) part of the application even though have different contents share the nature of procedural requirements.

Table 1. A comparison of judgement and application structures

Judgement structure	Application structure
(e) an account of the procedure followed	(e) a concise and legible statement confirming the applicant's compliance with the admissibility criteria laid down in Article 35 § 1 of the Convention
(f) the facts of the case	(e) a concise and legible statement of the facts
(g) a summary of the submissions of the parties	(f) statement of the alleged violation(s) of the Convention and...
(h) the reasons in point of law	... the relevant arguments [further details on the facts, alleged violations of the Convention and the relevant arguments]

---

[7] Art. 47 of the Rules 17 March 2022 https://www.echr.coe.int/documents/rules_court_eng.pdf "1. An application under Article 34 of the Convention shall be made on the application form provided by the Registry unless the Court decides otherwise. It shall contain all of the information requested in the relevant parts of the application form and set out (a) the name, date of birth, nationality and address of the applicant and, where the applicant is a legal person, the full name, date of incorporation or registration, the official registration number (if any) and the official address; (b) the name, address, telephone and fax numbers and e-mail address of the representative, if any; (c) where the applicant is represented, the dated and original signature of the applicant on the authority section of the application form; the original signature of the representative showing that he or she has agreed to act for the applicant must also be on the authority section of the application form; (d) the name of the Contracting Party or Parties against which the application is made; (e) a concise and legible statement of the facts; (f) a concise and legible statement of the alleged violation(s) of the Convention and the relevant arguments; and (g) a concise and legible statement confirming the applicant's compliance with the admissibility criteria laid down in Article 35 § 1 of the Convention. 2. (a) All of the information referred to in paragraph 1 (e) to (g) above that is set out in the relevant part of the application form should be sufficient to enable the Court to determine the nature and scope of the application without recourse to any other document. (b) The applicant may however supplement the information by appending to the application form further details on the facts, alleged violations of the Convention and the relevant arguments. Such information shall not exceed 20 pages.".

This structure and correspondence in court and applicants documents has been crucial in the well-known study published in 2016 by Aletras et al. which we have already discussed above.

5   Work in Progress in Italy

Now in Italy there is nothing which might be compared to the ECHR Rule 74. Nevertheless, some interesting early experiences in the use of a formalized structure of lawyers' and judges' briefs and judgments are worth signaling.

Between December 2015 and December 2017, the Directorate General for Information Services and Automation of the Ministry of Justice (DGSIA), promoted and funded a pilot project of collaboration between Milan judicial offices and the State Universities of Milan and Pavia to promote digitalization in court activity[8]. One line of research delved into the structure of judgments, both civil and criminal, concluding that the logical and legal structure of both types was essentially the same and that this constituted a great possibility for the development of models capable of opening up the application of AI techniques.

In the same span of time, the Superior Council of the Judiciary (CSM) and the National Forensic Council (CNF) signed in 2018 a Memorandum of understanding on clarity and conciseness in the drafting of pleadings and decisions on appeal. They shared the idea that the drafting of party pleadings and decisions should be inspired by the principles of conciseness and clarity. To this end, draft decision drafting schemes and criteria for drafting pleadings were developed and proposed to judges and lawyers. The next step, turning those drafts into templates entered by default into the Telematic Civil Process Console, has never happened[9].

Finally, the Person, Family and Juvenile Commission of the Milan Bar Association, in collaboration with Family Section of the Milan Civil Court, developed "templates for minutes and judgments to be prepared by the lawyers and to be filed in the telematic file"[10].

These few but significant experiences tell us how the time is now ripe for a decisive move beyond the models of the past and for a development of a digital by design process, forerunner of a digital by design law. Hereinafter we indicate the key milestones.

6   Toward a New-Generation Templates

August 2021, the Italian Ministry of Justice publishes a call reserved for Italian public universities, which, since its title, indicates the goal of technological innovation[11]. The

---

[8] Amedeo Santosuosso participated in that project as coordinator and Giulia Pinotti as an external researcher and PhD student (research topic: Digitization and public administration).

[9] A very embryonic attempt at Covid19 cause hearing postponement measures was carried out by a joint CSM-Ministry committee in 2020.

[10] More information at https://www.ordineavvocatimilano.it/it/commissione-persona-famiglia-e-minori/p14.

[11] "Finanziamento di interventi a regia nell'ambito dell'Asse I – Obiettivo Specifico 1.4 – Azione 1.4.1".

academic component that deals with the digitalization of judicial offices in Districts of the Courts of Appeal of Milan, Brescia, Turin, and Genoa, has focused on the need to take advantage of the existing digital data, to start using AI techniques and to boost a change in the way judges and lawyers write their documents (judgements and pleadings/briefs). This third point is twofold, as it starts creating data of better quality for future application of legal analytics on them and helps immediately the judicial assistants (Ufficio del processo) to learn what is the logical/legal structure of the documents, how to extract information from the materials and documents of cases and more.

The starting point is a conceptual template for the judicial assistants, which is divided into three parts (possibly marked in different colors): the first, concerns names of the parties and lawyers and other identifying elements such as background information about the litigation; the second part concerns the positions that the parties take in the trial and their claims; the third one implies the creation of the index of the document with the logical and legal order of the issues to be addressed and development of the actual motivation. The conceptual template actually consists of a plurality of models according to the stages of the judgment, the nature of the jurisdictional measure, and the type of proceeding. They should be shared with lawyers who should adopt similar and complementary conceptual templates of their pleadings and briefs as well. They can be technologically poor as they can be used on current word processing software.

The evolution of the conceptual template is given by new-generation templates, which are digitally native files, collecting structured data, included by default among the templates of Judges and Judicial Assistants, that are collected in appropriate data warehouse (DWH) and/or Data Lake (DL) and that can go into a dataset on which to operate with data retrieval tools, document builder and with AI techniques that enable knowledge extraction.

Briefly, the path of new-generation templates is important for (i) the efficiency of Judicial assistant offices (young inexpert graduated in law can immediately start working even on complex cases), (ii) lawyers who are called to improve the quality of their pleadings and briefs in terms of clarity, (iii) judges whose work can both improve its quality and be more productive and (iv), last but not least, the possibility, finally, to have a full development of legal analytics in the judicial activity.

6.1 An Example of Court of Appeal Formats

In this paragraph we show, in a very simplified way, what might be the structure of documents of judges and lawyers in a Court of appeal.

Table 2. A comparison of judgement and pleading formats

REPUBBLICA ITALIANA IN NOME DEL POPOLO ITALIANO Corte d'appello di xxxxxx	[CARTA INTESTATA AVVOCATO] Corte d'appello di xxxxxx
composta dai Signori: dott. Presidente dott. Relatore dott. Terzo giudice	
ha pronunciato la seguente SENTENZA nella causa civile n. __/__ R.G. promossa in grado d'appello	ATTO DI CITAZIONE IN APPELLO NELL'INTERESSE DI TIZIO per la riforma della Sentenza __ resa da ufficio giudiziario in data __ nella causa civile
TRA parte attrice Appellante E convenuto Appellato/appellante incidentale	TRA parte attrice Appellante E convenuto Appellato/appellante incidentale
Codice oggetto: oggetto	Codice oggetto: oggetto
LA SENTENZA DI PRIMO GRADO ha statuito	LA SENTENZA DI PRIMO GRADO ha statuito
MOTIVI di appello di TIZIO • TITOLO MOTIVO DI APPELLO A • TITOLO MOTIVO DI APPELLO B • TITOLO MOTIVO DI APPELLO C	INDICE DELL'ATTO • TITOLO MOTIVO DI APPELLO A • TITOLO MOTIVO DI APPELLO B • TITOLO MOTIVO DI APPELLO C
MOTIVI di appello incidentale di CAIO • TITOLO MOTIVO DI APPELLO INCIDENTALE D • TITOLO MOTIVO DI APPELLO INCIDENTALE E	
MOTIVI DELLA DECISIONE I punti sui quali la Corte è chiamata a pronunciarsi sono i seguenti: 1. .... 2. .... 3. ....n	
Su 1. Il giudice di prime cure ha statuito che xxxxxx L'appellante sostiene che ABSTRACT, motivo appello *** L'appellato contesta integralmente ritenendo che zzzzzz La Corte ritiene che il presente motivo di appello è fondato/ infondato e va accolto / respinto	MOTIVI DI APPELLO di TIZIO • PASSAGGIO DELLA SENTENZA IMPUGNATO • MOTIVO DI APPELLO A (titolo) • MOTIVO DI APPELLO A (esposizione) • MOTIVO DI APPELLO A ABSTRACT ***
Su 2. Il giudice di prime cure ha statuito che xxxxxx L'appellante sostiene che ABSTRACT, motivo appello *** L'appellato contesta integralmente ritenendo che zzzzzz La Corte ritiene che il presente motivo di appello è fondato/ infondato e va accolto / respinto	
Sulle spese di lite	Conclusioni e spese di lite
PQM La Corte, nella causa d'appello tra XXX e YYY così dispone: ZZZZZZZ	
Così deciso il ------ Il Consigliere estensore Il Presidente	Firma del difensore

Table 2 shows the similarity and complementarity in the formats for lawyers (on the right) and judges (on the left). The stress of the formats on the structure of the text, rather than the length per se, is noteworthy, and, on the other hand, the rigidity of the way in which the abstract of the defendant's reasons is automatically inserted into the text of the opinion of the Court. This serves to overcome a criticism that lawyers often direct at judges who, in their view, would not faithfully report their reasons.

Table 3 is a conceptual representation of the internal logical structure of the two formats, which is similar to Rule 74 (see Table 1) and different, for the automatic insertion of the abstract.

Table 3. A comparison of conceptual structures of judgements and pleadings

Court of Appeal	
Judgement structure	Pleading of lawyer structure
Heading *[Judicial office, Name of the judge(s), Parties and lawyers]*	Heading *[Judicial office, Name of the Appealing party and lawyer]*
Conclusions of the parties	Conclusions of the Appealing party
Background [the decision of first instance judge and other relevant information]	Background [the decision of first instance judge and other relevant information from the point of view of the party]
Index of the reasons of all parties in the appeal in the logical order given by the judges	
The reasons of the Court decision The Court replies to each reason of appeal reproducing the abstract given by the lawyer	Reasons for appeal Clearly divided for each appealed part of the first instance decision; each of them summarized at the end in a max 10 lines abstract

7   Towards a Document Builder

Once the dataset of legal materials is created according to a template of rules such as the one just outlined, a further step can be taken: a document builder, i.e. a tool to support the construction of new decisions, can be made available to judges. It is a system capable first of all of extracting information from the files of the case, but also of presenting the judge with judgments and snippets of reasoning, based on the applications by the parties, on similar cases that may support him in the decision.

The creation and use of document builder systems can represent a great development, in terms of both increased productivity and qualitative improvement (because they will be more informed decisions, about the state of the law and jurisprudence). At that point, the judge is left with the most difficult task, and one of the highest intellectual and professional value: to select that material and that proposed motivational path and challenge it, changing or specifying certain parameters, or certain elements of fact and law that contradict and change the consequentiality of the system's proposal.

Author Index

Audrito, Davide   102

Bermudez-Edo, Maria   25
Bernasconi, Anna   55
Billi, Marco   69

Cannataro, Mario   45
Cinaglia, Pietro   45
Costa, Mireia   15, 35, 55

Di Caro, Luigi   102

Fazio, Federica   112
Fernandes, Ana Xavier   5
Ferrara, Alfio   81
Ferreira, Filipa   5

García S., Alberto   15, 35, 55
Garcia-Moreno, Francisco M.   25
Garrido, José Luis   25
Guizzardi, Giancarlo   55

Lagioia, Francesca   69
León, Ana   5

Mármol, José Manuel Pérez   25

Pastor, Oscar   15, 35, 55
Picascia, Sergio   81
Pinotti, Giulia   112

Riva, Davide   81
Rodríguez-Fórtiz, María José   25
Rover, Aires José   69

Sabo, Isabela Cristina   69
Santin, Piera   102
Santos, Maribel Yasmina   5
Santosuosso, Amedeo   112
Sartor, Galileo   102
Sartor, Giovanni   69
Scherzinger, Stefanie   91
Storey, Veda C.   55
Sulis, Emilio   102

von Lewinski, Kai   91

Printed in the United States
by Baker & Taylor Publisher Services

Printed in the United States
by Baker & Taylor Publisher Services